LARSON, BOSWELL, KANOLD, STIFF

Passport
to Mathematics

BOOK 2

Answer Masters

Answer Masters includes short answers for each lesson to check homework. All answers are provided in the Teacher's Edition and the Complete Solutions Manual.

McDougal Littell
A HOUGHTON MIFFLIN COMPANY
Evanston, Illinois • Boston • Dallas

Answers for Lesson 1.1, pages 5–7

Ongoing Assessment
1. Yes   **2.** No   **3.** No

Practice and Problem Solving

1. b; answers vary.

2. 128

3. Answers vary.

4. *Sample answer:*

Width	Length	Perimeter	Pattern
1	2	6	
2	4	12	Perimeter is doubled.
3	6	18	Perimeter is tripled.
4	8	24	Perimeter is quadrupled.

5. White shark: 18 ft
Oceanic whitetip shark: 10 ft
Tiger shark: 15 ft

6. 2 units^2

7. 16

8. Running sneakers (r), walking sneakers (w),
aerobic sneakers (a), white (W), black (B), blue
laces (bl), red laces (rl), white laces (wl)
rWbl, rWrl, rWwl, rBbl, rBrl, rBwl, wWbl, wWrl,
wWwl, wBbl, wBrl, wBwl, aWbl, aWrl, aWwl,
aBbl, aBrl, aBwl

(continued)

Answers for Lesson 1.1, pages 5–7 (cont.)

9. 25–30 acres; multiply the number of acres needed for one household by 5.

10. A

11. D

12. 200; 0; two hundred is the largest number on the scale and zero is the least.

Answers for Lesson 1.2, pages 9–11

Ongoing Assessment
1. 820 partners

Practice and Problem Solving

1. 4, 9, 16. The sum is the square of the largest number being added; the sums increase by consecutive odd numbers.

2. 210

3. Answers vary.

4. 22; the numbers increase by 5.

5. 25; the numbers increase by 4, 5, 6, ...

6. 25; the numbers increase by 3, 5, 7, . . . ; or the numbers are squares of consecutive numbers.

7. 62; the numbers increase by 4, 8, 16, ...

8. 13; *Sample answer:* The number of triangles is 2 less than the number of sides.

9. *Sample answer:* Pattern is: 1, 1 + 2; 1 + 2 + 3, and so on; 20th number: 210.

10. 28; answers vary.

11. 304 ft; answers vary.

12. 435; answers vary.

13. 42; answers vary.

14. B

15. D

16. 1 ft or 12 in.; answers vary.

Answers for Spiral Review, page 12

1. 3.1

2. 16

3. 0.27

4.–6. *Sample answers*

4. $\frac{4}{10}$, $\frac{2}{5}$

5. $\frac{12}{16}$, $\frac{3}{4}$

6. $\frac{3}{9}$, $\frac{1}{3}$

7. 48

8. 3

9. 18

10. 15

Answers for Lesson 1.3, pages 16–19

Ongoing Assessment

1. Perimeter of tile: 20 in.
   Perimeter of larger diamond: 40 in.
   The perimeter of the large diamond is twice the perimeter of the tile.

2. Area of tile: 24 in.^{2}
   Area of larger diamond: 96 in.^{2}
   The area of the large diamond is four times the area of the tile because the large diamond consists of four tiles.

3. No

Practice and Problem Solving

1. Similar geometric figures have the same shape.

2. A and D; they have the same shape.

3. 44 cm, 105 cm^2

4. Answers vary.

5. 40 cm, 60 cm^2

6. Answers vary.

7.–10. Answers vary.

11. 32 m, 64 m^2

12. 34 ft, 60 ft^2

13. 30 ft, 30 ft^2

14. 24 in., 24 in.^{2}

15. 80 cm, 384 cm^2

16. 36 in., 93.6 in.^{2}

17. 56 in., 180 in.^{2}

18.

24 units^2

19. No. Answers vary.

20. Yes, same shape

(continued)

Answers for Lesson 1.3, pages 16–19 (cont.)

21. Yes, same shape

22. Yes, same shape

23. No, different shape

24. No, different shape

25. Yes, same shape

26. 36 cm, 54 cm^2

27.

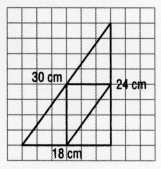

28. 72 cm, 216 cm^2

29. No. The perimeter of the tile is $\frac{1}{2}$ the perimeter of the similar triangle, while the area of the tile is $\frac{1}{4}$ the area of the similar triangle.

30. A

31. C

32. Answers vary.

Answers for Lesson 1.4, pages 21–24

Ongoing Assessment

1. Answers vary.

2. Least number of tosses is 6.

Practice and Problem Solving

1. 7; because it occurs in the table most often.

2. Answers vary.

3. About 8

4. Answers vary.

5. **a.** Answers vary but, since 1 in 4 of the paths takes the mouse to the food, you would expect the mouse to get there in 4 attempts.
   b. Simulations could include: Trace the path with a pencil. Flip a coin at each intersection; *heads* means "go left" and *tails* means "go right."

6. 45

7. Answers vary.

8. Simulations could include placing coins in a bag and removing one at a time until you have at least $.50.

9. Answers vary.

10. C

11. A

12. C

13. Answers vary.

Answers for Spiral Review, page 24

1. Five thousand, fifteen

2. Sixteen thousand, four hundred fifty

3. Four hundred twelve thousand, three

4. One hundred eighty thousand, nine hundred twenty-seven

5. 14.5

6. 61.2

7. 0.3

8. 798

9. 36 ft^2

10. 1.15 m^2

11. 54 mi^2

12. D

13. A

14. C

15. B

16. 22 mi/h

17. Cheetah

18. 42 mi/h

1. 24; 4

2. 4 sq units; the area is multiplied by 4; the area is multiplied by 9; the area is multiplied by 16; the area is multiplied by the square of the number that the side length is multiplied by; 100 sq units.

3. A and D; they have the same shape.

4. 42 in.^{2}

5. Answers vary.

6. 6 in.

7. 8 cm

8. 5 m

9. 40; $\frac{4}{5}$ of the spinner represents a free or discounted video, therefore, $\frac{4}{5} \times 50$ equals 40 videos.

10. 30; there are 36 possible combinations. Out of the 36 possibilities, 6 are doubles. $180 \times \frac{6}{36} = 30$ times.

11. Answers vary.

Answers for Lesson 1.5, pages 27–29

Ongoing Assessment

1. 4 balls in red region
   6 balls in blue region
   2 balls in yellow region

2. 5 balls in red region
   5 balls in blue region
   2 balls in yellow region

3. 1 ball in red region   or   4 balls in red region
   10 balls in blue region   3 balls in blue region
   1 ball in yellow region   5 balls in yellow region

Practice and Problem Solving

1. 36×36 is much bigger than 144.

2. 4×2 is much smaller than 36.

3.   764
   $+ 764$
   $\overline{1528}$

4.   476
       476
   $+ 476$
   $\overline{1428}$

5. Length $= 18$ m, width $= 1$ m

6. Length $= 13$ in., width $= 2$ in.

7. Length $= 5$ km, width $= 2$ km

8. Length $= 7$ ft, width $= 3$ ft

9. Length $= 11$ yd, width $= 3$ yd

10. Length $= 5$ mi, width $= 1$ mi

11.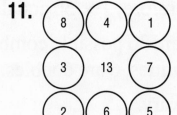

12.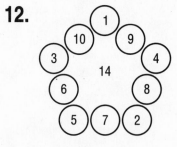

(continued)

Answers for Lesson 1.5, pages 27–29 (cont.)

13.

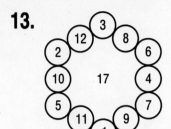

14. 17

15. 26

16. 45

17. 33 ft by 11 ft

18. 23, 23

19. 5

20. Z: 10, G: 2, P: 3

21. 25

22. D

23. B

24. 5 shelters that hold 35 people and 3 shelters that hold 64 people

Answers for Lesson 1.6, pages 31–34

Ongoing Assessment

1. $24 + n = 43$; answers vary.

2. $32 - n = 15$; answers vary.

3. $48 \div n = 16$; answers vary.

Practice and Problem Solving

1. Solving an equation means finding the value of the variable that makes the equation true. Answers vary.

2. Variable

3. Solution

4. Check

5. D

6. A

7. C

8. B

9. Number of CDs $= 7$,
   Cost of a CD (in dollars) $= c$,
   Total cost (in dollars) $= 83.93$; $7 \times c = 83.93$,
   $c = 11.99$; a CD costs $11.99.

10. No; 3

11. No; 8

12. Yes

13. Yes

14. No; 26

15. No; 6

16. 9

17. 25

18. 36

19. 35

(continued)

Answers for Lesson 1.6, pages 31–34 (cont.)

20. 7

21. 14

22. 11

23. 63

24. $b + 15 = 30, 15$

25. $d - 16 = 7, 23$

26. $4 \times n = 60, 15$

27. $t \div 18 = 2, 36$

28. 7.75

29. 19.03

30. 10.09

31. 13.2

32. 3.8

33. 13.65

34. Answers vary.

35. You get the original two-digit number.

36. 48 golf balls

37. $.45

38. A

39. C

40. C

41. Verbal Model:

Number of original reserves	+	Number of new reserves	=	Total number of reserves

Labels: Number of original reserves $= n$
Number of new reserves $= 14$
Total number of reserves $= 26$

Equation: $n + 14 = 26$

There were originally 12 reserves.

Answers for Spiral Review, page 34

1. A
2. C
3. D
4. B
5. 12

6. $256
7. 40 cm^2
8. 70 in.^{2}
9. 12 m^2
10. $15

Answers for Communicating About Mathematics, page 35

1. 448 pandas

2. 14 hours; 9 hours

3. 273,750 lbs

4. 1990: about 1290

5. **Verbal Model:**

$$\boxed{\begin{array}{c}\text{Hours}\\\text{of rest}\\\text{per day}\end{array}} \times \boxed{\begin{array}{c}\text{Days in}\\\text{one year}\end{array}} = \boxed{\begin{array}{c}\text{Hours of}\\\text{rest per}\\\text{year}\end{array}}$$

   Labels: Hours of rest per day = 9
   Days in one year = 365
   Hours of rest per year = x

   Equation: $9 \times 365 = x$

   Solution: 3285 hours

Answers for Lesson 1.7, pages 37–39

Ongoing Assessment

1. 19

2.

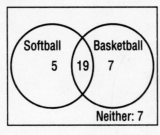

Softball 5 (19) Basketball 7
Neither: 7

Practice and Problem Solving

1. Center: Cindy; Guard: Shawna; Forward: Sheila

2. 8

3. Drama: Juan; Pep: Linda; Ski: Ed

4.

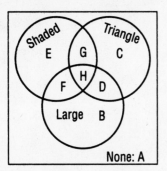

Shaded E G Triangle C
F H D
Large B
None: A

5.

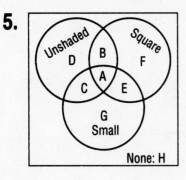

Unshaded D B Square F
C A E
G Small
None: H

6.

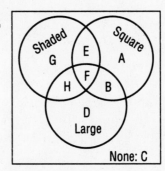

Shaded G E Square A
H F B
D Large
None: C

7. Ben; Jen

8. Len Su, Jen Lee, Ben Kipp, Ken Roi

9. 6

10. C

11. A

(continued)

12. 7 states

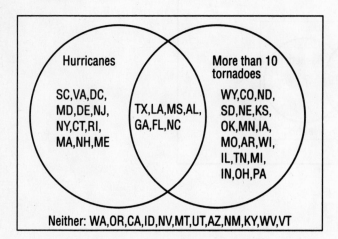

Hurricanes

SC,VA,DC,
MD,DE,NJ,
NY,CT,RI,
MA,NH,ME

TX,LA,MS,AL,
GA,FL,NC

More than 10
tornadoes

WY,CO,ND,
SD,NE,KS,
OK,MN,IA,
MO,AR,WI,
IL,TN,MI,
IN,OH,PA

Neither: WA,OR,CA,ID,NV,MT,UT,AZ,NM,KY,WV,VT

Answers for Lesson 1.8, pages 41–43

Ongoing Assessment

1. 91

2. Find the sum of the first 6 square numbers.

Practice and Problem Solving

1. 70

2. 20

3. 2 super and 4 large

4. 7

5. 28

6.–8. *Sample answers.* (Some intersections are numbered to show the order in which they are reached.)

6. 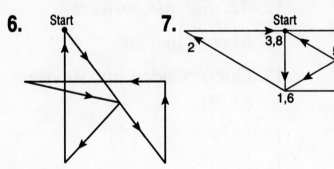   **7.**

8.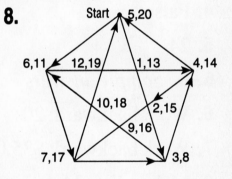

9. 24

10. 13

11. Yes, 4 ways without retracing

12. Minnows: one 1-lb and one 3-lb
Crayfish: one 3-pound

13. B

14. D

15. Tropical storms and hurricanes

Answers for Lesson 1.9, pages 45–47

Ongoing Assessment

1. About 366,000
2. About 280,000
3. About 2,238,000

Practice and Problem Solving

1.–2. *Sample answers*

1. Guess, Check, and Revise;
   Elephant: 13 ft
   Giraffe: 18 ft

2. Draw a diagram; 21

3. True

4. False

5. 24

6.–8. *Sample answers*

6. Work backward; 20

7. Work backward; $35.00

8. Make a list: 24

9. The greatest increase occurred in the morning hours. The afternoon temperatures increased slightly.

10. 11 A.M.

11. Between 2 P.M. and 3 P.M.

12.–13. *Sample answers*

12. Make a list; 20

13. Guess, Check, and Revise; $5.50

14. C

15. C

16. Seattle receives more rain, while Yuma has higher temperatures.

Answers for Chapter Review, pages 49–51

1. The Delicious apples are the better bargain. In the 3-pound bag, each pound costs $0.75.

2. Yes; there are 36 different meals.

3. Answers vary; 99

4. 48 m, 141.75 m^2

5. 6 cm square: 24 cm, 36 cm^2
   5 cm square: 20 cm, 25 cm^2

6. Answers vary.

7. 23, 33, 33

8. 27.4 m

9. 15

10. $m \times 9 = 63, 7$

11. A: 28, B: 140, C: 35, D: 60, F: 48, G: 42

12. Rod: tuna, Bill: cheese, Dan: chicken, Joe: ham and cheese

13. 29

14. Answers vary.

15. Multiply 68 by 5 and then subtract the four known readings.

Answers for Chapter Assessment, page 52

1. About 120

2. 55

3. **A** 12 ft, 6 ft^2
   B 20.8 m, 27.04 m^2

4. Arnold: dolphin
   Brenda: horse
   Chen: dog
   Dalia: cat

5. Nine 3-pound packages and five 4-pound packages

6. 18

7. $b + 44 = 98, 54$

8. $125 - a = 65, 60$

9. $n \times 12 = 60, 5$

10. $54 \div r = 3, 18$

11. One $10.00 ticket, two $1.40 tickets, one $.75 ticket

12. $18.00

Answers for Standardized Test Practice, page 53

1. C

2. D

3. B

4. A

5. A

6. D

Answers for Lesson 2.1, pages 59–61

Ongoing Assessment

1. Step 1: Multiply 6 by 4.
   Step 2: Add the result of Step 1 to 3.
   Step 3: Divide the result of Step 2 by 3.
   Step 4: Multiply 4 by 2.
   Step 5: Subtract the result of Step 4 from the result of Step 3.

2. Steps should follow order of operations.

Practice and Problem Solving

1. A numerical expression has only numbers and symbols; $4 \times 3 + 7$. A variable expression has at least one variable; $(12 + n) \times 8$.

2. First do operations within grouping symbols. Then multiply and divide from left to right. Finally add and subtract from left to right.

3. **a.** The total cost to make n pins
   b. The total amount received for selling n pins
   c. The total profit from selling n pins
   d. The total profit from selling 1 pin
   e. The total profit from selling n pins

4. Answers vary.

5. Addition was performed before division.
$$12 + 6 \div 2 = 12 + 3$$
$$= 15$$

(continued)

Answers for Lesson 2.1, pages 59–61 (cont.)

6. Addition was performed before multiplication.

$$3 + 4(7 - 5) = 3 + 4 \times 2$$
$$= 3 + 8$$
$$= 11$$

7. 16

8. 10

9. 19

10. 8

11. 6

12. 4

13. 12

14. 8

15. 36

16. 2

17. 15

18. 64

19. True

20. False; $(6 + 10) \div 2 = 8$

21. False; $9 + 4 - (3 + 7) = 3$

22. True

23. False; $15 \div (5 - 2) + 14 = 19$

24. False; $(8 + 16) \div (2 \times 2) = 6$

25. True

26. True

27. False; $(7 + 2) \div (7 - 4) = 3$

28. $12 \times 90 \div 6$; 180

29. $14 + 6 \times 3$; 32

30. $15.7 - 14.25 \div 5$; 12.85

31. $25.2 - 2.8 \times 3$; 16.8

32. 5, 8, 11, 14; numbers increase by 3.

33. 40, 28, 18, 10; numbers decrease by 12, 10, and 8.

34. $n \times (n + 1)$; 110

35. B

36. B

37.
- Locate a site.
- Survey and map the site.
- Use heavy machinery to move large amounts of soil.
- Use picks and brushes to remove soil from artifacts.
- Photograph and record locations where artifacts were found.

Answers for Lesson 2.2, pages 65–67

Ongoing Assessment

1. 16-by-16-by-16

2. *Sample answer:* Guess, check, and revise

Practice and Problem Solving

1. 8^2

2. 4^3

3. B

4. A

5. D

6. C

7. $2 \times 3 = 6$ and
$2^3 = 2 \times 2 \times 2 = 8$

8. $3 + 5^2 = 3 + 5 \cdot 5$
$\qquad = 3 + 25 = 28$
and
$(3 + 5)^2 = 8^2$
$\qquad = 8 \cdot 8 = 64$

9. c^3

10. 6^3

11. m^5

12. 2^6

13. 81

14. 100

15. 125

16. 1

17. 64

18. 49

19. 1

20. 81

21. 32

22. B

23. 4, 16, 64, 256, 1024, 4096;
each power is 4 times the
previous power.

24. 64; 512; 4096; 32,768;
262,144; 2,097,152;
each power is 8 times the
previous power.

25. a. No, $4^2 = 16$, not 8.
b. No, $16 + 4 \div 2 = 16 + 2$

26. 18, 10

27. 0, 26

28. 2, 6

(continued)

Answers for Lesson 2.2, pages 65–67 (cont.)

29. 4, 16

30. 2, 10

31. 8, 56

32. 61

33. $<$

34. $=$

35. $<$

36. $=$

37. $<$

38. $=$

39. Each number in the first row is 2 more than the previous number. Each number in the second row is 3 more than the previous number. The numbers in each row increase by 1 more than the numbers in the previous row increased by.

40. 1, 8, 27, 64, 125, 216, 343; 1^3, 2^3, 3^3, 4^3, 5^3, 6^3, 7^3; the consecutive whole numbers cubed starting with 1.

41. D

42. C

43. 4; $3^4 \times 4 + 3^3 + 1 = 81 \times 4 + 27 + 1$
$$= 324 + 27 + 1$$
$$= 352$$

Answers for Spiral Review, page 68

1. 96; each number is two times the preceding number.

2. 17; the numbers increase by the pattern 1, 3, 5, 7, 9, ...

3. 27; each number is the preceding number divided by 3.

4. 32; each number is 8 more than the preceding number.

5. 2 quarters, 2 dimes, 3 nickels

6. Yes

7. No, $t = 37$

8. No, $d = 6$

9. 16

10. 13

11. 51

12. 45

Answers for Lesson 2.3, pages 71–73

Ongoing Assessment

1. If 6 people join, the band has 126 members. 126 is evenly divisible by 1, 2, 3, 6, 7, and 9. There are six possible rectangular formations.

2. Possible sketches include rectangles of 1-by-126, 2-by-63, 3-by-42, 6-by-21, 7-by-18, and 9-by-14.

Practice and Problem Solving

1. 360. The smallest number that is divisible is 180. The number between 300 and 400 is 180 + 180 or 360.

2. 3

3. Yes. The last digit of a number that is divisible by 10 is 0. Any number whose last digit is 0 is divisible by 5.

4. The numbers are divisible as follows.

    30: by 2, 3, 5, 6, 10   984: by 2, 3, 4, 6
    54: by 2, 3, 6, 9   1806: by 2, 3, 6
    164: by 2, 4   4760: by 2, 4, 5, 10
    235: by 5   9478: by 2
    576: by 2, 3, 4, 6, 9   15,507: by 3, 9
    723: by 3   25,390: by 2, 5, 10

5. Always; since 6 is even, any number divisible by 6 must be even and thus divisible by 2.

6. Sometimes. 15 is divisible by 5 and not by 4; 20 is divisible by both 5 and 4.

(continued)

Answers for Lesson 2.3, pages 71–73 (cont.)

7. Sometimes. 9 is divisible by 3 and not by 6; 12 is divisible by both 3 and 6.

8. Divisible by 3

9. Divisible by none

10. Divisible by all

11. Divisible by 2, 3, 4, 6

12. Divisible by 3, 5, 9

13. Divisible by 2

14. Divisible by 2, 3, 6, 9

15. Divisible by 2, 4, 5, 10

16. 2, 5, 8

17. 0, 6

18. 2, 8

19. 0, 3, 6, 9

20. *Sample answer:* 130

21. *Sample answer:* 1035

22. *Sample answer:* 4311

23. Yes; *Sample answer:* $2^4 = 16, 2^3 = 8$, and $16 \div 8 = 2$.

24. 90; *Sample answer:* $6 = 2 \times 3, 9 = 3 \times 3$, so the smallest number is $2 \times 3 \times 3 \times 5 = 90$.

25. The number formed by its last three digits is divisible by 8.

26. 100 one dollar bills,
20 five dollar bills,
10 ten dollar bills,
5 twenty dollar bills,
2 fifty dollar bills,
1 one hundred dollar bill

27. No

28. A

29. D

30. 1152 squares with dimensions 1 foot-by-1 foot
288 squares with dimensions 2 feet-by-2 feet
128 squares with dimensions 3 feet-by-3 feet
72 squares with dimensions 4 feet-by-4 feet
32 squares with dimensions 6 feet-by-6 feet
18 squares with dimensions 8 feet-by-8 feet
8 squares with dimensions 12 feet-by-12 feet
2 squares with dimensions 24 feet-by-24 feet

Answers for Lesson 2.4, pages 77–80

Ongoing Assessment

1.
$22 = 3 + 19$	$36 = 5 + 31$	$50 = 7 + 43$
$24 = 5 + 19$	$38 = 7 + 31$	$52 = 11 + 41$
$26 = 3 + 23$	$40 = 3 + 37$	$54 = 13 + 41$
$28 = 5 + 23$	$42 = 11 + 31$	$56 = 3 + 53$
$30 = 7 + 23$	$44 = 13 + 31$	$58 = 5 + 53$
$32 = 13 + 19$	$46 = 17 + 29$	$60 = 7 + 53$
$34 = 5 + 29$	$48 = 19 + 29$	

2. The list does not prove that *every* even number greater than 2 can be written as the sum of two primes.

Practice and Problem Solving

1.

```
              84
            /    \
         2  ×  42
              /    \
           2 × 6 × 7
          /    |    \
       2 × 2 × 3 × 7
```

$2^2 \cdot 3 \cdot 7$

2. 1

3. 2

4. Multiply the factors to get the original number.

5. Composite

6. Prime factorization

7. Neither: 1; Prime: 2, 3, 5, 7, 11, 13, 17, 19, 23;
Composite: 4, 6, 8, 9, 10, 12, 14, 15, 16, 18, 20, 21, 22, 24, 25

8. $2^4 \cdot 3$

9. $2^3 \cdot 7$

10. $2 \cdot 3 \cdot 5^2$

11. $2^4 \cdot 3^2$

12. $2^5 \cdot 3 \cdot 7$

13. $2^4 \cdot 3^2 \cdot 5$

14. $3 \cdot 5 \cdot 11^2$

15. $2 \cdot 3 \cdot 5 \cdot 7 \cdot 11$

16. 66

17. 105

(continued)

18. 56

19. 54

20. 84

21. 90

22. 11; prime

23. 26; composite

24. 41; prime

25. 1287 is divisible by 3 because $1 + 2 + 8 + 7 = 18$ and 18 is divisible by 3.

26.
$7 = 2 + 2 + 3$
$9 = 3 + 3 + 3$
$11 = 3 + 3 + 5$
$13 = 3 + 5 + 5$
$15 = 5 + 5 + 5$
$17 = 5 + 5 + 7$
$19 = 5 + 7 + 7$
$21 = 7 + 7 + 7$
$23 = 5 + 5 + 13$
$25 = 5 + 7 + 13$
$27 = 7 + 7 + 13$
$29 = 3 + 13 + 13$

No. The list does not prove that every odd number can be written as the sum of three prime numbers.

27. 1-by-195, 13-by-15, 3-by-65; *Sample answer:* 5 rows of 39

28. 35

29. 36

30. 18 or 24

31.

Height		Width
2 feet	by	2 feet
2 feet	by	3 feet
2 feet	by	4 feet
2 feet	by	6 feet
2 feet	by	12 feet
4 feet	by	2 feet
4 feet	by	3 feet
4 feet	by	4 feet
4 feet	by	6 feet
4 feet	by	12 feet
8 feet	by	2 feet
8 feet	by	3 feet
8 feet	by	4 feet
8 feet	by	6 feet
8 feet	by	12 feet

32. B

33. D

34. C

35. Answers vary.
$2 \cdot 2 \cdot 7 = 28$
$15 \Rightarrow$ ˧·χ.

Not enough information to write 26 because $26 = 2 \cdot 13$ and the symbol for 13 is not given.
$33 \Rightarrow$ ˧·⌡.

Answers for Spiral Review, page 80

1.

1 ▭ 45

3 ▭ 15

5 ▭ 9

2.

1 ▭ 48

2 ▭ 24

3 ▭ 16

4 ▭ 12

6 ▭ 8

3.

1 ▭ 72

2 ▭ 36   6 ▭ 12

3 ▭ 24   8 ▭ 9

4 ▭ 18

4.

1 ▭ 80

2 ▭ 40   5 ▭ 16

4 ▭ 20   8 ▭ 10

5. 27

6. 110

7. 57

8. $52

9. E

10. G

11. C

12. B

13. F

14. D

15. 39; 78

16. 25; 22

17. 101; 104

18. 48; 24

Answers for Mid-Chapter Assessment, page 81

1. $5 \times (9 + 2) = 55$
2. $45 - 6 \times (2 + 3) = 15$
3. $36 \div (3 + 6) \div 2 = 2$
4. $(27 - 18) \div (6 - 3) = 3$
5. 125
6. 64
7. 243
8. 64
9. C
10. A
11. D
12. B
13. 1-by-84, 2-by-42, 3-by-28, 4-by-21, 6-by-14, 7-by-12
14. 130
15. 132
16. 225
17. 252
18. False; $2^3 = 2 \cdot 2 \cdot 2 = 8$
19. True
20. False; The expression $6 \times 8 - 6 \div 2$ is equal to 45.
21. True
22. Yes, 3 is a factor of both 36 and 78.

Answers for Lesson 2.5, pages 83–85

Ongoing Assessment

1. 6; List the factors of 24, 30, and 54. From the list, 6 is the greatest common factor.

2. 4; List the factors of 16, 52, and 80. From the list, 4 is the greatest common factor.

3. 8; List the factors of 32, 40, and 72. From the list, 8 is the greatest common factor.

Practice and Problem Solving

1. 1, itself

2. No; the common prime factors are 2 and 3, so the greatest common factor is $2 \times 3 = 6$.

3. a. Factors of 32: 1, 2, 4, 8, 16, 32;
Factors of 48: 1, 2, 3, 4, 6, 8, 12, 16, 24, 48;
Greatest common factor: 16

b. $32 = 2^5$; $48 = 2^4 \cdot 3$;
Greatest common factor: $2^4 = 16$

c. Answers vary.

4. Cut the subs into 6-inch pieces; No, 3-inch pieces

5. B	**9.** 7	**13.** 44
6. D	**10.** 1	**14.** 15
7. C	**11.** 12	**15.** 9
8. A	**12.** 6	**16.** 8

(continued)

Answers for Lesson 2.5, pages 83–85 (cont.)

17.–20. *Sample answer:* Multiply the given number by 2 different primes to get the pair.

17. $15, 21; 3 \times 5 = 15, 3 \times 7 = 21$

18. $33, 55; 11 \times 3 = 33, 11 \times 5 = 55$

19. $24, 60; 12 \times 2 = 24, 12 \times 5 = 60$

20. $32, 48; 16 \times 2 = 32, 16 \times 3 = 48$

21. Sometimes. *Sample answer:* Greatest common factor of 7 and 14 is 7, greatest common factor of 12 and 15 is 3.

22. Sometimes. *Sample answer:* Greatest common factor of 8 and 9 is 1, greatest common factor of 12 and 15 is 3.

23. a. 1:1, 2:2, 3:3, 4:2, 5:1
6:6, 7:1, 8:2, 9:3, 10:2
11:1, 12:6, 13:1, 14:2
15:3, 16:2, 17:1, 18:6

b.

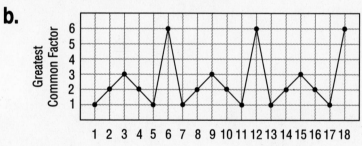

c. The pattern 1, 2, 3, 2, 1, 6 repeats.

24. D

25. C

26. a. 6   **b.** 12   **c.** 6

Answers for Lesson 2.6, pages 87–90

Practice and Problem Solving

1. B

2. C

3. A

4. A, C; because the greatest common factor of their numerator and denominator is 1.

5. B, D, E

6. No answer required.

7.–10. *Sample answer:* multiply 3 by some prime number that is not a factor of the numerator.

7. $\frac{12}{21}$

8. $\frac{30}{33}$

9. $\frac{42}{51}$

10. $\frac{93}{69}$

11. $\frac{2}{3}$

12. $\frac{1}{10}$

13. $\frac{3}{7}$

14. $\frac{3}{4}$

15. $\frac{7}{11}$

16. $\frac{11}{30}$

17. $\frac{1}{6}$

18. $\frac{5}{17}$

19. $\frac{3}{5}$

20. $\frac{14}{28}, \frac{28}{56}, \frac{7}{14}$

21. $\frac{72}{81}, \frac{8}{9}, \frac{24}{27}$

22. $\frac{24}{28}, \frac{12}{14}, \frac{6}{7}$

23. $\frac{15}{25}, \frac{75}{125}, \frac{3}{5}$

24. $\frac{50}{200}, \frac{5}{20}$, simplest form: $\frac{1}{4}$

25. $\frac{54}{63}, \frac{36}{42}$, simplest form: $\frac{6}{7}$

26. $\frac{64}{72}, \frac{16}{18}$, simplest form: $\frac{8}{9}$

(continued)

Answers for Lesson 2.6, pages 87–90 (cont.)

27. $\frac{3}{8}$

28. $\frac{1}{36}$

29. $\frac{1}{6}$

30. $\frac{1}{9}$

31. $\frac{19}{72}$

32. $\frac{1}{18}$

33. B

34. D

35. A

36. Answers vary.

Answers for Spiral Review, page 90

1. 16

2. 10

3. 1 head and 1 tail;
answers vary.

4. $6.50

5. $(22 - 10) \times (16 \div 8) = 24$

6. 169

7. 125

8. 16

9. 49

10. $2^3 \cdot 3^2 \cdot 5$

11. $3^3 \cdot 5^2$

12. $2 \cdot 3 \cdot 5 \cdot 7^2$

13. $7 \cdot 13^2$

14. 36

Answers for Communicating About Mathematics, page 91

1. Field goal: 3 points, safety and conversion: 2 points

2. Yes; 13 is a factor of 48,841.

3. The height of a scoop of ice cream is about 2 inches.

4. No; 28 is not a factor of 48,841 because 2 is a factor of 28 but not a factor of 48,841.

Answers for Lesson 2.7, pages 95–97

Ongoing Assessment

1. $\frac{1}{9}$, answers vary.

2. $\frac{2}{7}$, answers vary.

3. $\frac{1}{6}$, answers vary.

Practice and Problem Solving

1. Always

2. Sometimes

3. Sometimes

4. Answers vary.

5. *Sample answer:*

1 muffin

Each $\bigtriangleup$ = $\frac{1}{4}$ muffin

Each person will get $10 \div 4$ or $10 \times \frac{1}{4} = 2\frac{1}{2}$.

6. C

7. A

8. B

9. $\frac{4}{5}$; 0.8

10. $\frac{17}{20}$; 0.85

11. $\frac{6}{8}$; 0.75

12. $\frac{24}{50}$; 0.48

13. 0.67

14. 0.63

15. 0.78

16. 0.92

17. $500, $100, $50: each $\frac{2}{27}$ or 0.07; $20: $\frac{6}{27}$ or 0.22; $10, $5, $1: each $\frac{5}{27}$ or 0.19

(continued)

18. 0.35

19. 0.19

20. 0.34

21. 0.35

22. $0.\overline{18}, 0.\overline{27}, 0.\overline{36}, 0.\overline{45}, 0.\overline{54}, 0.\overline{63}, 0.\overline{72}, 0.\overline{81}, 0.\overline{90}$;
sample answer: the repeating digits are the product of the numerator and 9.

23. $0.\overline{1}, 0.\overline{2}, 0.\overline{3}, 0.\overline{4}, 0.\overline{5}, 0.\overline{6}, 0.\overline{7}, 0.\overline{8}$; *sample answer:* the repeating digit is the numerator.

24. $\frac{1}{20}$

25. $\frac{1}{10}$

26. $\frac{1}{8}$

27. $\frac{1}{25}$

28. $\frac{13}{49} \approx 0.27$; $\frac{1}{11} \approx 0.09$; $\frac{1}{15} \approx 0.07$; $\frac{1}{9} \approx 0.11$

29. C

30. B

31. $\frac{32}{65}$; 0.49

Ongoing Assessment

1. Answers vary.

2. Answers vary.

Practice and Problem Solving

1. *Sample answer:* ordering finish times in a race; ordering wrenches in a tool set

2. $a = \frac{7}{10}, b = \frac{23}{20}$ or $1\frac{3}{20}; \frac{7}{10} < \frac{23}{20}$

3. False

4.

$\frac{1}{3}, \frac{3}{4}, \frac{4}{5}, \frac{6}{5}, 1.4, 2.05$

5.

$\frac{5}{4}, \frac{8}{3}, 2.85, 2.9, 3.3, \frac{7}{2}$

6. Marin; Marin ran 3.75 miles, which is farther than Sheila (3.4 miles) and you (3.6 miles) ran.

7. $a = \frac{1}{4}, b = \frac{18}{10}$ or $1\frac{4}{5}, \frac{1}{4} < \frac{18}{10}$

8. $a = \frac{79}{20}$ or $3\frac{19}{20}, b = \frac{87}{20}$ or $4\frac{7}{20}; \frac{79}{20} < \frac{87}{20}$

(continued)

Answers for Lesson 2.8, pages 99–101 (cont.)

9.

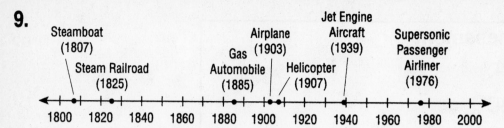

10.

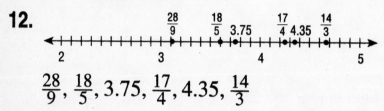

5.4, 5.45, 6.75, 6.8, 7.05, 7.5

11.

$\frac{7}{9}, \frac{8}{7}, \frac{7}{3}, \frac{18}{6}, \frac{13}{4}, \frac{7}{2}$

12.

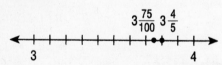

$\frac{28}{9}, \frac{18}{5}, 3.75, \frac{17}{4}, 4.35, \frac{14}{3}$

13. True

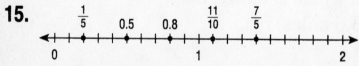

14. False

15.

Each number is $\frac{3}{10}$ more than the preceding number. 1.7 or $\frac{17}{10}$, 2

(continued)

16.

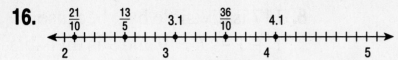

Each number is 0.5 more than the preceding number. 4.6 or $\frac{23}{5}$, 5.1 or $\frac{51}{10}$

17. B

18. A

19.

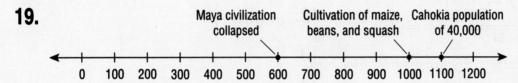

Answers for Chapter Review, pages 103–105

1. 6

2. 22

3. $24 \div (8 - 2) + 6 \cdot 3 = 22$

4. $>$

5. 12

6. Divisible by 2, 3, 4, 6

7.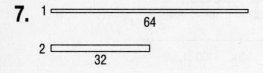

8. 117 is divisible by 3 because $1 + 1 + 7 = 9$ and 9 is divisible by 3.

9. $2 \cdot 3 \cdot 5^3$

10. 32

11. 2×2

12. $\frac{5}{8}$, $\frac{30}{48}$, $\frac{70}{112}$

13. $\frac{6}{7}$

14. 0.2

15. 0.12

16. 0.09

17.

$$\frac{9}{11}, 0.85, \frac{8}{9}, \frac{9}{10}, 0.95$$

18. *Sample answer:* $\frac{7}{25}$

Answers for Chapter Test, page 106

1. 16

2. 8

3. 7

4. 21

5. 32

6. 31

7. 2, 3, 6, and 9

8. all

9. 3

10. $2 \times 3 \times 13$

11. $2^4 \times 7$

12. $3^3 \times 5$

13. 8

14. 5

15. 22

16. $\frac{1}{9}$

17. $\frac{2}{3}$

18. $\frac{4}{5}$

19. 0.4

20. 0.38

21. 0.78

22.

$\frac{1}{5}, \frac{21}{20}, \frac{7}{5}, \frac{3}{2}, \frac{11}{6}, 2.35$

23.

$\frac{13}{4}, \frac{10}{3}, 3.85, \frac{29}{7}, \frac{9}{2}, 5.2$

24.

Answers for Standardized Test Practice, page 107

1. B

2. A

3. D

4. C

5. B

6. D

7. B

8. D

9. B

10. A

Answers for Lesson 3.1, pages 115–117

Practice and Problem Solving

1. False, change 3 to 60.

2. True

3. True

4. False, change 240 to 60.

5. *Sample answers:* 6 and 504, 18 and 168, 24 and 126, or 42 and 72.

6. 30,000

7. 14

8. 18

9. 20

10. 84

11. 24

12. 224

13. 230

14. 280

15. 576

16. 30

17. $23,088, $23,244, $23,400, $23,556, $23,712, $23,868

18. 4, 6; 12

19. 6, 9; 18

20. 10, 15; 30

21. $2 \cdot 3 \cdot n$ or $6 \cdot n$

22. The sailboat pattern. The sailboat pattern can occur a whole number of times on *each* wall, while the grape pattern cannot.

23. Yes, at 18 feet and 36 feet.

24. B

25. C

26. 13

Answers for Lesson 3.2, pages 119–121

Ongoing Assessment

1. Answers vary.

2. Answers vary.

Practice and Problem Solving

1. A fraction is proper if it is less than 1 such as $\frac{5}{6}$. A fraction is improper if it is greater than or equal to 1 such as $\frac{9}{7}$.

2. $3\frac{1}{5}$, not $1\frac{3}{5}$

3. $8\frac{2}{3} = 8 + \frac{2}{3}$
   $= \frac{24}{3} + \frac{2}{3}$
   $= \frac{26}{3}$, not $\frac{10}{3}$

4. $3\frac{3}{4}$; $\frac{15}{4}$

5. 2.75 lb

6. Proper, less than 1

7. Improper, equal to 1

8. Improper, greater than 1

9. Proper, less than 1

10. Yes, see Exercise 7.

11. D

12. A

13. C

14. B

15. $1\frac{5}{6}$

16. $6\frac{3}{4}$

17. $3\frac{1}{9}$

18. $4\frac{3}{8}$

19. $\frac{16}{11}$

20. $\frac{20}{7}$

21. $\frac{25}{3}$

22. $\frac{21}{2}$

23. 4.25

24. 1.6

25. 9.5

26. 1.15

27. **a.** $\frac{7}{4}$; $\frac{5}{4}$ **b.** 7; 5

28. D

29. D

(continued)

Answers for Lesson 3.2, pages 119–121 (cont.)

30. a. $13.75; $13.88; $13.50; $13.38; $13.25

    b. Tuesday

    c. Friday

Answers for Spiral Review, page 122

1. 10 mi

2. True

3. False, $20 \div (2 \times 5) + 7 = 9$

4. False, $(28 - 16) \div (2 + 4) = 2$

5. 1

6. 34

7. 3

8. 27

9. $\frac{2}{9}$

10. $\frac{1}{4}$

11. $\frac{5}{7}$

12. $\frac{7}{8}$

Answers for Lesson 3.3, pages 127–129

Ongoing Assessment

1. Yes

$$\frac{1}{6} + \frac{3}{8} + \frac{1}{8} + \frac{1}{3} = \frac{8}{48} + \frac{18}{48} + \frac{6}{48} + \frac{16}{48} \qquad \text{Use common denominator of 48.}$$

$$= \frac{8 + 18 + 6 + 16}{48} \qquad \text{Add numerators.}$$

$$= \frac{48}{48} \qquad \text{Simplify.}$$

$$= 1$$

Practice and Problem Solving

1. 24; find the least common multiple of 12 and 8.

2. Common denominator: write the sum or difference of the numerator over the denominator. Different denominator: rewrite fractions with a common denominator and then add or subtract.

3. Your friend, $\frac{1}{10}$ mi

4. Yes. The student has used a common denominator, though the common denominator (32) that was used was not the least common denominator (8).

5. $1\frac{2}{7}$ or $\frac{9}{7}$

6. $\frac{2}{3}$

7. $\frac{23}{35}$

8. $\frac{13}{15}$

9. $\frac{1}{10}$

10. $\frac{11}{24}$

11. $1\frac{1}{8}$

12. $\frac{19}{20}$

13. Yes; *sample answer:*
$\frac{3}{4} + \frac{3}{4} = 1\frac{1}{2}$

(continued)

14. $\frac{5}{18}, \frac{4}{15}, \frac{3}{12}; \frac{2}{9}, \frac{1}{6}$

15. $\frac{7}{10}, \frac{9}{10}, 1\frac{1}{10}; 1\frac{3}{10}, 1\frac{5}{10}$

16. $1\frac{1}{4}, 1\frac{1}{6}, 1\frac{1}{8}; 1\frac{1}{10}, 1\frac{1}{12}$

17. $\frac{1}{4}$

18. $\frac{2}{11}$

19. $\frac{1}{3}$

20. $\frac{4}{7}$

21. $\frac{1}{6}$

22. $\frac{3}{4}$

23. $\frac{14}{25}$

24. $\frac{4}{5}$

25. $\frac{31}{100}$

27. D

28. B

29. $\frac{43}{60}; \frac{23}{60}$

Answers for Lesson 3.4, pages 131–134

Ongoing Assessment

1. $5\frac{1}{2}$; find common denominator and simplify.

2. $5\frac{1}{3}$

3. $4\frac{5}{6}$; find common denominator and regroup.

Practice and Problem Solving

1. C, A, D, B

2. D, C, B, A

3. 11

4. 4

5. 13

6. Answers vary.

7. $5\frac{4}{5}$

8. $6\frac{3}{4}$

9. $2\frac{2}{11}$

10. $3\frac{1}{2}$

11. 3

12. 13

13. 9

14. 5

15. $8\frac{1}{9}$

16. $8\frac{7}{12}$

17. $1\frac{4}{5}$

18. $4\frac{3}{4}$

19. 7

20. $7\frac{19}{24}$

21. $1\frac{3}{8}$

22. $4\frac{13}{16}$

23. $3\frac{2}{3}$

24. $9\frac{1}{9}$

25. $4\frac{5}{12}$

26. $8\frac{23}{24}$

27. $9\frac{1}{5}$

28. $1\frac{1}{3}$

29. $2\frac{1}{4}$

30. $5\frac{7}{24}$

31. $6\frac{1}{3}$ ft

32. 5 ft

33. $5\frac{7}{12}$ ft

34. $1\frac{2}{5}$

35. 19 h

36. $4\frac{1}{2}$ inches

37. $3\frac{3}{4}$ inches

38. $6\frac{1}{2}$ inches

39. C

40. a. $3\frac{3}{8}$

   b. $16\frac{1}{8}$; 11

   c. Higher, $3\frac{7}{8}$

Answers for Spiral Review, page 134

1. 4 sq units
2. 44 triangles, 10 squares
3. 3 m by 6 m
4. 5 in. by 9 in.
5. 72 in.^{2}
6. 125
7. 256

8. 121
9. $2 \times 2 \times 2 \times 2 \times 2 \times 2$
10. $2 \times 2 \times 2 \times 3 \times 3$
11. $3 \times 5 \times 7$
12. 1×47
13. $2 \times 2 \times 2 \times 2 \times 5$
14. $2 \times 2 \times 2 \times 3 \times 3 \times 3$

Answers for Mid-Chapter Assessment, page 135

1. 42
2. 68
3. 110
4. 60
5. Every 12 weeks
6. B
7. D
8. C
9. A

10. $\frac{4}{25}$
11. $\frac{39}{100}$
12. $\frac{9}{100}$
13. 1
14. $1\frac{7}{9}$
15. $6\frac{3}{5}$
16. $4\frac{4}{9}$
17. $9\frac{11}{24}$

18. $3\frac{1}{10}$
19. $2\frac{1}{4}$
20. $5\frac{17}{24}$
21. $\frac{2}{5}, 0.4$
22. $\frac{2}{3}, 0.\overline{6}$
23. $\frac{1}{2}, 0.5$
24. $\frac{1}{2}, 0.5$

Answers for Lesson 3.5, pages 139–141

Ongoing Assessment

1. $\frac{1}{2}$

2. $3\frac{3}{5}$

3. $2\frac{4}{15}$

4. $4\frac{7}{12}$

Practice and Problem Solving

1. $\frac{5}{6}$ of the squares has been shaded blue and $2\frac{1}{3}$ of the squares have been shaded yellow; the area of the two-color region is $1\frac{17}{18}$. So, the picture is a model for the equation $\frac{5}{6} \times 2\frac{1}{3} = 1\frac{17}{18}$.

2. $\frac{3}{4}$ of the square has been shaded blue and $\frac{2}{5}$ of the square has been shaded yellow; the area of the two-color region is $\frac{6}{20}$. So, the picture is a model for the equation $\frac{3}{4} \times \frac{2}{5} = \frac{6}{20}$.

3. First rewrite as improper fractions. Then multiply numerators and denominators.

4. $\dfrac{a}{b} \cdot \dfrac{c}{d} = \dfrac{a \cdot c}{b \cdot d}$

5. Either way is OK.

6. Mercury: 40 lb, Venus: 88 lb, Earth: 100 lb, Mars: 38 lb, Jupiter: 250 lb, Saturn: 105 lb, Uranus: 90 lb, Neptune: 114 lb, Pluto: 5 lb

7. always

8. sometimes

9. never

10. $\frac{3}{7}$

11. $\frac{10}{27}$

12. $\frac{24}{5}$ or $4\frac{4}{5}$

13. $\frac{5}{2}$ or $2\frac{1}{2}$

14. $\frac{14}{3}$ or $4\frac{2}{3}$

(continued)

Answers for Lesson 3.5, pages 139–141 (cont.)

15. $\frac{99}{8}$ or $12\frac{3}{8}$

16. $\frac{37}{63}$

17. $\frac{19}{18}$ or $1\frac{1}{18}$

18. 22

19. 13

20. $\frac{2}{3}$

21. $8\frac{2}{5}$

22. $1\frac{17}{64}$ in.^{2}

23. $7\frac{1}{3}$ ft^2

24. $\frac{1}{3}$ mi

25. $4\frac{3}{8}$ m^2

26. D

27. D

28. a. $9\frac{1}{4}$ yd **b.** Yes

Answers for Lesson 3.6, pages 143–46

Ongoing Assessment

1. Method 2 uses the Distributive Property.

2. Answers vary.

Practice and Problem Solving

1. $3 \times (6 + 8) = 3 \times 6 + 3 \times 8$

2. $(5 + 2) \times 3 = 5 \times 3 + 2 \times 3$

3. Error: Adding $3 + 3$.
$$3(10) + 3(15) = 3(10 + 15)$$
$$= 3(25)$$
$$= 75$$

4. Error: Not multiplying $6 \times \frac{5}{8}$.
$$6\left(1 + \frac{5}{8}\right) = 6(1) + 6\left(\frac{5}{8}\right)$$
$$= 6 + \frac{30}{8}$$
$$= 6 + 3\frac{6}{8}$$
$$= 9\frac{6}{8} \text{ or } 9\frac{3}{4}$$

5. $5 \times t, 5 \cdot t, 5(t), 5t$

6. Only c

7. $5 \times 4 + 5 \times 3; 5(4 + 3)$

8. $3 \times 8 + 1 \times 8; (3 + 1)8$

9. 55

10. 1700

11. 71

12. $4\frac{1}{3}$

13. $3\frac{1}{5}$

14. 10

15. $3t + 3 \cdot 4$

16. $12 \cdot 5 + 12b$

17. $4c + 6c$ or $10c$

18. $15n + 25n$

19. $1x + 1 \cdot 24$

20. $5p + 5q$

21. 48, 48; the expressions are equal.

22. 22, 22; the expressions are equal.

23. 33.6

24. 100.5

25. 125

26. 22.33

27. 97

28. 36.12

(continued)

Answers for Lesson 3.6, pages 143–46 (cont.)

29. 150 min or $2\frac{1}{2}$ hr

30. No, three packages weigh 57 lb.

31. $200.50

32. C

33. B

34. B

35. $921\frac{1}{2}$ gallons

Answers for Spiral Review, page 146

1. 15

2. 20

3. 9

4. 2

5. 9 people

6. Divisible by 3

7. Divisible by 2, 3, 6

8. Divisible by all

9. 17

10. 2

11. 1

12. 3

13. 35

14. 4

15. $\frac{1}{2}$

16. $\frac{7}{12}$

17. $\frac{3}{4}$

18. $\frac{1}{28}$

19. $\frac{2}{5}$

20. $\frac{3}{4}$

21. $\frac{2}{5}$

22. $\frac{2}{3}$

23. $\frac{4}{5}$

24.

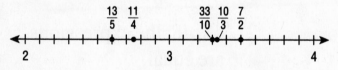

25.

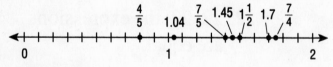

26. Friday

Answers for Communicating About Mathematics, page 147

1. The transatlantic flight in 1928 took $7\frac{1}{6}$ hours or 7 hours 10 minutes longer than in 1932.

2. About 3 times longer. Airplane travel is much faster today.

3. $\frac{5}{48}$; 10

4. $473\frac{1}{2}$ miles; answers vary.

Answers for Lesson 3.7, pages 151–153

Ongoing Assessment

1. Yes **2.** No **3.** Yes **4.** No

Practice and Problem Solving

1. $2\frac{1}{2} \div \frac{5}{6}$, 3

2. Error: Multiplying the second fraction by the reciprocal of the first, instead of multiplying the first fraction by the reciprocal of the second.

$$\frac{3}{7} \div \frac{1}{5} = \frac{3}{7} \times \frac{5}{1}$$
$$= \frac{15}{7} \text{ or } 2\frac{1}{7}$$

3. Rewrite $1\frac{3}{4}$ and $2\frac{5}{8}$ as improper fractions, multiply $\frac{7}{4}$ by the reciprocal of $\frac{21}{8}$, multiply 7 times 8 and mutliply 4 times 21, simplify.

4. Rewrite $3\frac{1}{2}$ as an improper fraction, multiply $\frac{7}{2}$ by the reciprocal of $\frac{1}{10}$, multiply 7 times 10 and write the result over 2, simplify.

5. Yes, 10

6. B

7. D

8. C

9. A

10. $\frac{1}{6}$

11. $1\frac{24}{25}$

12. 20

13. $\frac{4}{11}$

14. $\frac{3}{7}$

15. $\frac{56}{39}$ or $1\frac{17}{39}$

16. 12

17. $\frac{3}{4}$

18. $\frac{5}{9}$

19. $\frac{1}{5}$

20. $\frac{3}{8}$

21. $\frac{6}{4}$ or $\frac{3}{2}$ or $1\frac{1}{2}$

22. $\frac{1}{7}$

23. $\frac{8}{3}$ or $2\frac{2}{3}$

24. $6\frac{2}{5}$ or 6 whole servings

25. $2\frac{1}{24}$ feet; $1\frac{1}{2}$ times longer

26. A

27. C

28. Answers vary; $27.25

1. 18

2. 30

3. 176

4. 360

5. $6\frac{4}{5}$

6. $\frac{17}{2}$

7. 4.375

8. $\frac{22}{9}$

9. $2\frac{5}{12}$ ft

10. $2\frac{7}{10}$ m

11. $7\frac{1}{24}$

12. $6\frac{13}{56}$

13. $5\frac{3}{4}$

14. $\frac{13}{15}$

15. $6\frac{9}{10}$ million

16. $\frac{1}{2}$ million

17. $\frac{7}{12}$

18. $20\frac{44}{45}$ m^2

19. 22.4

20. $6m + 12$; 30

21. $\frac{5}{6}$

22. $9\frac{23}{24}$ inches

Answers for Chapter Assessment, page 158

1. 90

2. 210

3. 204

4. 360

5.

6. $\frac{1}{3}$

7. $5\frac{2}{7}$

8. $11\frac{1}{6}$

9. $2\frac{3}{20}$

10. $\frac{17}{18}$

11. $\frac{1}{2}$

12. $2\frac{5}{8}$

13. $1\frac{17}{24}$

14. $10\frac{1}{12}$

15. $\frac{7}{10}$

16. $1\frac{5}{7}$

17. $52\frac{1}{2}$

18. 4

19. $2\frac{1}{27}$

20. 50

21. $1\frac{1}{18}$

22. $\frac{1}{6}$

23. $2\frac{1}{2}$

24. $2m + 12$

25. $8t + 40$

26. $12p + 23p$

27. Downhill

28. $\frac{5}{24}$

29. Cross country: $\frac{1}{6} \times 48 = 8$
Downhill: $\frac{3}{8} \times 48 = 18$
Water: $\frac{1}{4} \times 48 = 12$
Jet: $\frac{1}{12} \times 48 = 4$

30. Don't ski: $\frac{1}{8} \times 48 = 6$

Answers for Standardized Test Practice, page 159

1. B

2. C

3. C

4. B

5. D

6. B

7. C

8. A

Answers for Cumulative Review, pages 160 and 161

1. *Sample answer:* separate into cases, make a list, solve a simpler problem, look for a pattern.

2. Not similar

3. Similar

4. Similar

5. 8 m by 10 m

6. 9 in. by 16 in.

7. $102 - n = 54; 48$

8. $15x = 75; 5$

9. $n \div 8 = 9; 72$

10. 5

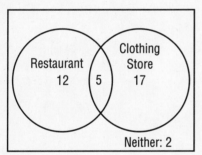

11. 17

12. 29

13. 2

14. 16

15. 1

16. 216

17. False; change *a number* to *an even number.*

18. False; change 3 to 9 and change 9 to 3.

19. True

20. True

21. False; change 5 to 25.

22. True

23. $\frac{1}{10}$

24. $\frac{1}{2}$

25. $\frac{5}{7}$

26. $\frac{11}{12}$

27. $\frac{1}{6}$

28. $\frac{2}{3}$

29. 0.38

30. 0.42

31. 0.17

32. 0.22

33. 0.2

34. 0.35

35.

$$\frac{2}{9}, \frac{1}{4}, \frac{3}{11}, \frac{2}{5}$$

(continued)

Answers for Cumulative Review, pages 160 and 161 (cont.)

36.

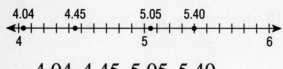

4.04, 4.45, 5.05, 5.40

37.

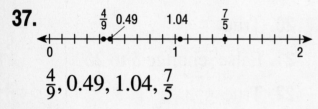

$\frac{4}{9}$, 0.49, 1.04, $\frac{7}{5}$

38. 60

39. 77

40. 408

41. 120

42. 60

43. 80

44. $16t + 96$

45. $\frac{11}{25}$

46. Less than half an hour; *Sample answer:* this section takes up more than half the circle.

47. $\frac{13}{20}$

48. $\frac{41}{60}$

Answers for Lesson 4.1, pages 169–171

Ongoing Assessment

1.

Number of Swims	*Computation*	*Total Cost*
0	4(0)	$0
1	4(1)	$4
2	4(2)	$8
3	4(3)	$12
4	4(4)	$16
5	4(5)	$20
6	4(6)	$24
7	4(7)	$28
8	4(8)	$32

2. Total cost $= 4n$, where n is the number of times you swim in a month.

3.

Times Swam	0	1	2	3	4	5	6	7	8
Example 1 ($)	10	12	14	16	18	20	22	24	26
Example 2 ($)	5	8	11	14	17	20	23	26	29
Youth Club ($)	0	4	8	12	16	20	24	28	32

4. If you swim 4 times or less in a month, the youth club is the least expensive. If you swim 5 times in a month, all three cost the same.

Practice and Problem Solving

1. 1, 6, 11, 16, 21

2. 20, 19.7, 19.4, 19.1, 18.8

3. A

4. C

5. B

6. 45, 90, 135, 180, 225; $45n$

7.

x	1	2	3	4	5	6
$3x - 1$	2	5	8	11	14	17

(continued)

8.

x	1	2	3	4	5	6
$x \div 3$	$\frac{1}{3}$	$\frac{2}{3}$	1	$1\frac{1}{3}$	$1\frac{2}{3}$	2

9.

x	1	2	3	4	5	6
$4 + 5x$	9	14	19	24	29	34

10.

x	1	2	3	4	5	6
$33 - 2x$	31	29	27	25	23	21

11.

x	1	2	3	4	5	6
$7 + 4x$	11	15	19	23	27	31

12.

x	1	2	3	4	5	6
$3x \div 4$	$\frac{3}{4}$	$1\frac{1}{2}$	$2\frac{1}{4}$	3	$3\frac{3}{4}$	$4\frac{1}{2}$

13.

m	1	2	3	4	5
$3m + 2$	5	8	11	14	17

14.

m	1	2	3	4	5
$16 - 2m$	14	12	10	8	6

15.

m	1	2	3	4	5
$10m - 10$	0	10	20	30	40

16. As n increases by 1, time decreases by 3.

17. As n increases by 1, time increases by 6.

18. $3.90 + 0.10n$

19. When $n < 7$, the first company does; because, for $n < 7$, it charges less. When $n > 7$, the second company does; because, for $n > 7$, it charges less.

20. B

21. D

22.

Week	1	2	3	4	5
Laps	4	6	8	10	12

$2w + 2$

Answers for Lesson 4.2, pages 173–175

Ongoing Assessment

1.

x	1	2	3	4	5	6	7
y	1	3	5	7	9	11	13

2.

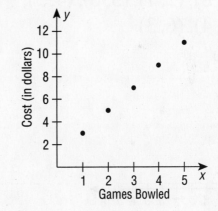

Practice and Problem Solving

1. C

2. B

3. D

4. E

5. A

6. F

7. $(0, 0)$

8. 3, 5, 7, 9, 11, 13

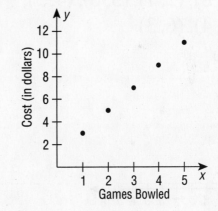

(continued)

9. (0, 3)

10. (2, 5)

11. (0, 0)

12. (2, 0)

13. (3, 3)

14. (5, 4)

15. (1, 2), (2, 4), (3, 6), (4, 8), (5, 10), (6, 12)

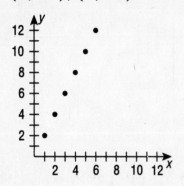

16. (1, 8), (2, 7), (3, 6), (4, 5), (5, 4), (6, 3)

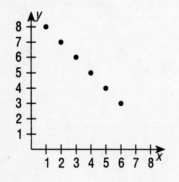

17. 4, 8, 12, 16, 20

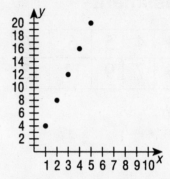

For every increase of x by 1, y increases by 4.

18. $1\frac{1}{2}, 2\frac{1}{2}, 3\frac{1}{2}, 4\frac{1}{2}, 5\frac{1}{2}$

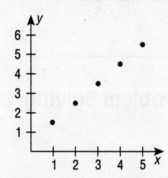

For every increase in x by 1, y increases by 1.

19. 28.5, 27, 25.5, 24, 22.5

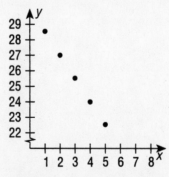

For every increase in x by 1, y decreases by 1.5.

(continued)

Answers for Lesson 4.2, pages 173–175 (cont.)

20. Company A: 35, 45, 55, 65, 75, 85

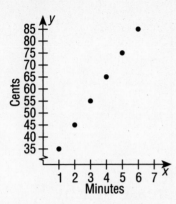

Company B: 38, 46, 54, 62, 70, 78

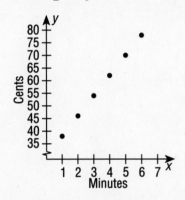

Company A is less expensive for calls that are 2 minutes or less. Company B is less expensive for calls that are more than 2 minutes long.

21.

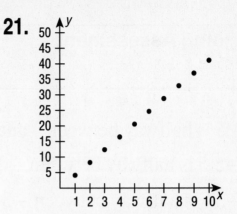

For every increase in x by 1, y increases by 4.1.

22. C

23. B

24.

The team won when they scored 8 or more runs in a game.

Answers for Spiral Review, page 176

1. 2, 4, 8, 16; the number of regions doubles each time the paper is folded.

2. 5

3. 28

4. 14

5. 14

6. 11

7. 4

8. >

9. >

10. >

11. >

12. 3 mi; $\frac{9}{16}$ mi^2

13. $17\frac{4}{5}$ ft, $18\frac{7}{10}$ ft^2

14. 8 yd, $2\frac{2}{3}$ yd^2

Answers for Lesson 4.3, pages 179–181

Ongoing Assessment

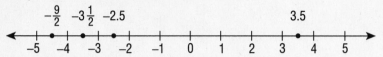

1. 3.5 is halfway between 3 and 4.

2. -2.5 is halfway between -2 and -3.

3. $-3\frac{1}{2}$ is halfway between -3 and -4.

4. $-\frac{9}{2}$ is halfway between -4 and -5.

Practice and Problem Solving

1. $3.5,\ -6\frac{1}{2},\ -2.75$

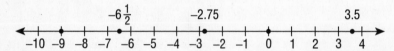

2. a: -5; b: -3; c: -1; d: 1; e: 7   **4.** $3 > -1$

3. Zero   **5.** $-2 > -5$

6.

Numbers increase by 4; 11, 15.

7.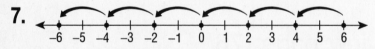

Numbers decrease by 2; -4, -6.

8.

Numbers in odd-numbered positions increase by 1, while numbers in even-numbered positions decrease by 1; 3, -2.

(continued)

Answers for Lesson 4.3, pages 179–181 (cont.)

9.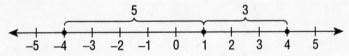

Numbers increase by 1, by 2, by 3, and so on; 7, 13.

10. <

11. >

12. >

13. <

14. False

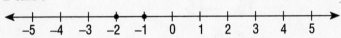

15. False

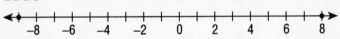

16. True

17. −12

18. 5

19. −45

20. −9

21. −9, −8, −3, 0, 3

22. −2.5, −$\frac{9}{4}$, −$\frac{1}{2}$, 2, 5.5

23. −7, −5, −4.25, −$\frac{7}{2}$, −1$\frac{3}{4}$

24. −6.8, −2.3, −2$\frac{1}{5}$, $\frac{3}{8}$

25. Sole

26. 1800 ft; 3300 ft; 6000 ft; 15,000 ft; 36,000 ft

27. B

28. A

29. D

30. Buttons: 20, Banners: −18, Magnets: −15, Programs: 30, Pencils: −3, Hats: 8; −18, −15, −3, 8, 20, 30; Profit

Answers for Lesson 4.4, pages 185–188

Ongoing Assessment

1. & 2. The order of addition does not affect the answer. The Commutative Property of Addition is illustrated.

1. 1   **2.** −7

Practice and Problem Solving

1. B

2. C

3. A

4. never

5. sometimes

6. always

7. Change $\boxed{-}$ to $\boxed{+/-}$.

8. Insert $\boxed{+}$ before 6.

9. $-6 + 5, -1$

10. $1 + (-4), -3$

11. 13

12. 0

13. 7

14. 11

15. −15

16. 0

17. 23

18. −13

19. C; −1

20. B; −5

21. A; 1

22. To the integers in the odd position, add 1. To the integers in the even position, add −2.
$-2, -1$

23. To the integers in the odd position, add −6. To the integers in the even position, add 4.
$-9, -5$

(continued)

Answers for Lesson 4.4, pages 185–188 (cont.)

24. 1

25. 5

26. 2

27. 0

28. Positive

29. Zero

30. Negative

31. −4

32. 4

33. 0

34. −915 ft

35. −367 ft

36. 0; Answers vary.

37. A

38. A

39. Chris P., +8; Pat S., +9; Fran B., +11; Sam F., +12
Lowest score: Chris P.
Order: Chris P., Pat S., Fran B., Sam F.

Answers for Spiral Review, page 188

1. Three 3-pound bags, one 5-pound bag

2. 5

3. 36

4. 16

5. 2, 5, 8

6. 1, 4, 7

7. 2, 5, 8

8. 0, 3, 6, 9

9. $\frac{1}{3}$

10. 70

11. 5

12. $5 \cdot 8 + 5 \cdot 4$, 60

13. $\frac{1}{4} \cdot 8 + \frac{1}{4} \cdot 6$, $3\frac{1}{2}$

14. $3\left(\frac{2}{9} + \frac{1}{9}\right)$, 1

15. $2(3.2 + 4.8)$, 16

Answers for Mid-Chapter Assessment, page 189

1.

x	1	2	3	4	5	6
$3x + 8$	11	14	17	20	23	26

2.

x	1	2	3	4	5	6
$100 - 5x$	95	90	85	80	75	70

3.

x	1	2	3	4	5	6
$120 \div x$	120	60	40	30	24	20

4. When x increases by 1, the cost increases by 5.

5. When x increases by 1, the cost decreases by 7.

6. $(2, 6), (3, 5), (4, 4), (5, 3), (6, 2), (7, 1)$

7. $\left(1, 2\frac{1}{2}\right), (2, 3), \left(3, 3\frac{1}{2}\right), (4, 4), \left(5, 5\frac{1}{2}\right), (6, 6)$

8.

x	1	2	3	4	5	6
y	1.5	2.5	3.5	4.5	5.5	6.5

$(1, 1.5), (2, 2.5), (3, 3.5), (4, 4.5), (5, 5.5), (6, 6.5)$

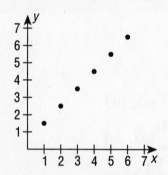

As x increases by 1, y increases by 1.

(continued)

9.

x	1	2	3	4	5	6
y	$6\frac{1}{2}$	$5\frac{1}{2}$	$4\frac{1}{2}$	$3\frac{1}{2}$	$2\frac{1}{2}$	$1\frac{1}{2}$

$$\left(1, 6\tfrac{1}{2}\right), \left(2, 5\tfrac{1}{2}\right), \left(3, 4\tfrac{1}{2}\right), \left(4, 3\tfrac{1}{2}\right), \left(5, 2\tfrac{1}{2}\right), \left(6, 1\tfrac{1}{2}\right)$$

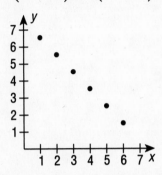

As x increases by 1, y decreases by 1.

10.

x	1	2	3	4	5	6
y	0	1	2	3	4	5

$(1, 0), (2, 1), (3, 2), (4, 3), (5, 4), (6, 5)$

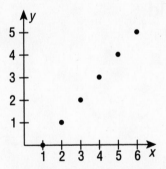

As x increases by 1, y decreases by 1.

11. $-10, -5, -1, 4$

12. $-15, -13, -12, -11$

13. -11

14. 4

15.

1st mi	2nd mi	3rd mi	4th mi
4	-3	0	-2

$4 + (-3) + 0 + (-2) = -1$, so his time for the whole race was 1 second less.

Answers for Lesson 4.5, pages 193–195

Practice and Problem Solving

1. Negative

2. Negative

3. Positive

4. Negative

5. -10%

6. -18%

7. When you subtract a positive integer, you should move to the left.

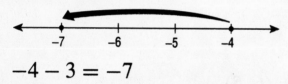

$-4 - 3 = -7$

8. When you subtract a negative integer, you should move to the right.

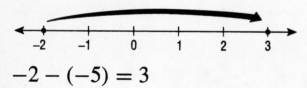

$-2 - (-5) = 3$

9. B

10. C

11. A

12. -9

13. -19

14. 19

15. 17

16. 2

17. -11

18. -15

19. -48

20. 62

(continued)

Answers for Lesson 4.5, pages 193–195 (cont.)

21. -101

22. -183

23. 111

24. 247

25. 35

26. -30

27. -34

28. 0

29. 382

30. -12

31. $2\frac{3}{4}$ in.

32. $\frac{7}{8}$ in.

33. $2\frac{1}{2}$ in.

34. D

35. B

36. $-100 - (-30) = -70$ feet

Answers for Lesson 4.6, pages 199–202

Practice and Problem Solving

1. Subtract 12 from both sides.

2. Subtract 7 from both sides.

3. B

4. A

5. C

6. No, solutions are -9 and 1.

7. Yes, both solutions are 6.

8. **Verbal Model:** $\boxed{\text{Previous balance}} + \boxed{\text{Deposit}} = \boxed{\text{New balance}}$

   Labels: Previous balance $= a$
   Deposit $= \$65$
   New balance $= \$315$

   Equation: $a + 65 = 315$

   Solution: $\$250$

9. -14

10. -13

11. -23

12. -9

13. -50

14. -39

15. 0.9

16. $\frac{1}{3}$

17. $\frac{1}{2}$

18. $1119 + x = 1458$, $x = 339$ thousand

19. 1991, 1994

20. *Sample answer:*
    $m + 6 = 0$;
    $35 + p = 29$;
    $r + 3 = -3$

21. $x + 10 = 10, 0$

22. $15 + x = -15, -30$

(continued)

23. $-20 = 14 + x, -34$

24. $26 = x + 17, 9$

25. $32 + t = 212, t = 180° \, \text{F}$

26. Answers vary.

27. Verbal Model: $\boxed{\begin{array}{c}\text{Cost} \\ \text{of CDs}\end{array}} + \boxed{\begin{array}{c}\text{Cost} \\ \text{of tapes}\end{array}} = \boxed{\begin{array}{c}\text{Money} \\ \text{you have}\end{array}}$

Labels: Cost of CDs $= c$
Cost of cage $= \$8.75$
Money you have $= \$28$

Equation: $c + 8.75 = 28$

Solution: You can spend up to $19.25, so you could buy two CD singles or an album CD.

28. D

29. C

30. A

31. a. Michigan

b. Michigan, Notre Dame, Alabama, Texas, Nebraska, Penn State, Ohio State, Oklahoma, Tennessee, Southern Cal. No, the table list is by winning percentage.

Answers for Spiral Review, page 202

1. 34 mm; 48 mm^2

2. Ashley is 16 years old. Carmen is 11 years old. Felicia is 8 years old. Ricardo is 14 years old.

3. 3^4

4. $2 \cdot 3^2 \cdot 5$

5. $2^4 \cdot 7$

6. $3 \cdot 5 \cdot 11$

7. $\frac{3}{5}$

8. 8

9. $\frac{8}{9}$

10. $\frac{17}{300}$

11. D

12. A

13. B

14. C

15. -3

16. 55

17. 10

18. -5

Answers for Communicating About Mathematics, page 203

1.–2.

2. $-12,200$
The Titanic is $-13,000 - (-800)$ or $-12,200$ feet deeper.

3. -180 feet

4. 3282 years; $1982 - (-1300) = 3282$

5. -32 feet

Answers for Lesson 4.7, pages 205–207

Ongoing Assessment

1. -8; to get x by itself, subtract 21 from each side.

2. $-\frac{1}{2}$; to get n by itself, add $\frac{1}{2}$ to each side.

3. 18; to get m by itself, add 4 to each side.

4. -9.8; to get p by itself, subtract 2.5 from each side.

Practice and Problem Solving

1. Subtract 11 from both sides.

2. Add 7 to both sides.

3. Subtract -15 or add 15 to both sides.

4. Subtract 13 from both sides.

5. A

6. D

7. B

8. C

9. 65

10. Answers vary.

11. Negative, -8

12. Negative, -6

13. Positive, 23

14. *Sample answer:*
   $z + 3 = -6; z - 4 = -13$

15. *Sample answer:*
   $-9 = b + 7; -20 = b - 4$

16. *Sample answer:*
   $12 + k = 24; 10 - k = -2$

17. 0

18. 16

19. 34

20. -8

21. -23

22. 11

23. 3.33

24. 25.52

25. $3\frac{1}{2}$

26. $x - 15 = -15, 0$

27. $26 = x + 40, -14$

28. $\frac{7}{8} = x - \frac{1}{4}, \frac{9}{8}$

29. Yes; each solution is 8.

30. $x - 14\frac{3}{4} = 12\frac{1}{2}, 27\frac{1}{4}$ feet

(continued)

31. 3

32. $t - 13 = 25; t = 38°$ F

33. Verbal Model: $\boxed{\text{Original price}} - \boxed{\text{Discount}} = \boxed{\text{Sale price}}$

Labels: Original price $= p$
Discount $= \$9$
Sale price $= \$27$

Equation: $p - 9 = 27$

Solution: $\$36$

34. Verbal Model: $\boxed{\text{Sale price}} + \boxed{\text{Tax}} = \boxed{\text{Total cost}}$

Labels: Sale price $= \$27$
Tax $= t$
Total cost $= \$28.35$

Equation: $27 + t = 28.35$

Solution: $\$1.35$

35. C

36. C

37. Verbal Model: $\boxed{\begin{array}{c}\text{Your team's} \\ \text{finish}\end{array}} = \boxed{\begin{array}{c}\text{1st place} \\ \text{team's finish}\end{array}} + \boxed{\begin{array}{c}\text{Time} \\ \text{difference}\end{array}}$

Labels: Your team's finish $= 3.35$ minutes
1st place team's finish $= x$ minutes
Time difference $= 0.15$ minute

Equation: $3.35 = x + 0.15$

Solution: 3.2 minutes

Verbal Model: $\boxed{\begin{array}{c}\text{Your team's} \\ \text{finish}\end{array}} = \boxed{\begin{array}{c}\text{3rd place} \\ \text{team's finish}\end{array}} - \boxed{\begin{array}{c}\text{Time} \\ \text{difference}\end{array}}$

Labels: Your team's finish $= 3.35$ minutes
3rd place team's finish $= x$ minutes
Time difference $= 0.05$ minute

Equation: $3.35 = x - 0.05$

Solution: 3.4 minutes

Answers for Chapter Review, pages 209–211

1.

n	1	2	3	4	5	6
$n+4$	5	6	7	8	9	10

2.

n	1	2	3	4	5	6
$6-2n$	4	2	0	-2	-4	-6

3.

n	1	2	3	4	5	6
$3n-13$	-10	-7	-4	-1	2	5

4.

n	1	2	3	4	5	6
$-n+3$	2	1	0	-1	-2	-3

5. The value of the expression be-
gins at 23. Each time t increases
by 1, the value of the expression
decreases by 1.
$23-t$

6. $A(3, 1)$

7. $B(4, 3)$

8. $C(1, 2)$

9. $(1, 5), (2, 4), (3, 2), (4, 3),$
$(5, 1)$

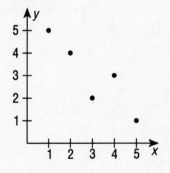

10. $-7, -6, -2, 1, 6$

11. $-70, -54, -48, -46,$
$-45, -35, -34, -32$

12. Montana; Connecticut

13. -17

14. -4

15. 0

16. -5

17. Yes

18. 6

19. 16

20. -4

21. -5

22. -7

23. 26

24. 8

25. $x + 3.25 = 8$, $4.75

26. 19

27. 0

28. -17

Answers for Chapter Assessment, page 212

1.

t	0	1	2	3	4	5	6
$2t + 6$	6	8	10	12	14	16	18

2.

t	0	1	2	3	4	5	6
$t \div 4$	0	$\frac{1}{4}$	$\frac{1}{2}$	$\frac{3}{4}$	1	$1\frac{1}{4}$	$1\frac{1}{2}$

3.

t	0	1	2	3	4	5	6
$15 - 3t$	15	12	9	6	3	0	-3

4. 3, 6, 9, 12, 15

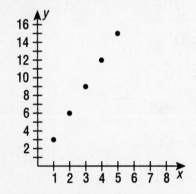

As x increases by 1,
y increases by 3.

5. $\frac{1}{2}$, $1\frac{1}{2}$, $2\frac{1}{2}$, $3\frac{1}{2}$, $4\frac{1}{2}$

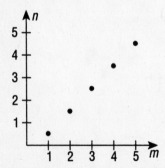

As m increases by 1,
n increases by 1.

6. 15, 12, 9, 6, 3

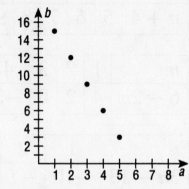

As a increases by 1,
b decreases by 3.

7. $>$

8. $<$

9. $>$

10. 7

11. 14

12. -19

13. -28

14. 19

15. -7

16. -9

17. -36

18. 0

19. 57

20. $\frac{7}{12}$

21. 4.55

(continued)

Answers for Chapter Assessment, page 212 (cont.)

22. $40 + 0.75n$; answers vary.

23. Labels: Sale price $= 63.75$ dollars

                    Discount $= d$ dollars

                    Original price $= 85$ dollars

    Equation: $63.75 + d = 85$

    Solution: $21.25

Answers for Standardized Test Practice, page 213

1. C

2. B

3. D

4. B

5. A

6. A

7. C

8. A

Answers for Lesson 5.1, pages 221–223

Ongoing Assessment

1. 38

2. 5 or greater

3. 2

Practice and Problem Solving

1. The median is 15.5, not 15.

2. There is no mode.

3. The median is 14, not 16.

4. 55

5. 120

6. 77

7. $3.38, $3.25, $3.25

8. $3\frac{7}{15}, 4, 4$

9. 4, 4, 3

10. $1\frac{1}{2}, 1, 0$

11. 8.24, 8, 8; 8; the mean, median, and mode are all about 8.

12. *Sample answer:* 8, 8, 10, 11, 13

13. *Sample answer:* 25, 25, 25, 25, 30

14. $39,000; $27,000; $25,000 and $27,000. The median, because most of the employees have salaries near $27,000.

15. The mean would be higher ($41,457.14), the median and the mode would remain the same.

16. C

17. D

18. *Sample answer:* the median is 1.8, because it is between the mode of 1.7 and the mean of 1.9.

Answers for Lesson 5.2, pages 227–229

Ongoing Assessment

Interval	Tally	Frequency
0 - 1		0
2 - 3		0
4 - 5		0
6 - 7		0
8 - 9	III	3
10 - 11		0
12 - 13	I	1
14 - 15	III	3
16 - 17	JHT I	6
18 - 19	JHT JHT I	11
20 - 21	JHT JHT	10
22 - 23	JHT II	7
24 - 25	IIII	4
26 - 27	III	3
28 - 29	I	1
29 - 30	I	1

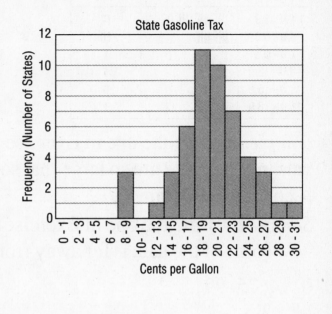

1. *Sample answer:* more vertical bars

2. Advantage: more information with smaller intervals
   Disadvantage: takes longer to create graph

Practice and Problem Solving

1.

Interval	Tally	Frequency
1.0 - 1.9	IIII	4
2.0 - 2.9	JHT III	8
3.0 - 3.9	JHT	5
4.0 - 4.9	III	3
5.0 - 5.9	JHT II	7
6.0 & up	III	3

(continued)

2.

Interval	Tally	Frequency
1.0 - 3.9	JHT JHT JHT II	17
4.0 - 6.9	JHT JHT	10
7.0 - 9.9	I	1
10.0 - 12.9	I	1
13.0 - 15.9		0
16.0 - 18.9	I	1

Sample answer: the one in Exercise 1; the intervals in Exercise 2 are so large that some of the data is condensed too much.

Sample answer: the one in Exercise 2; it shows how some of the data is far away from the rest.

3. 0–14; 75–99

4. Answers vary.

5. *Sample answer:*

Interval	Tally	Frequency
0 - 9	JHT IIII	9
10 - 18	JHT JHT	10
19 - 27	JHT JHT I	11
28 - 36	JHT I	6
37 - 45	III	3
46 - 54	IIII	4
55 - 63		0
64 - 72	III	3
73 - 81		0
82 - 90	III	3
91 - 99	I	1

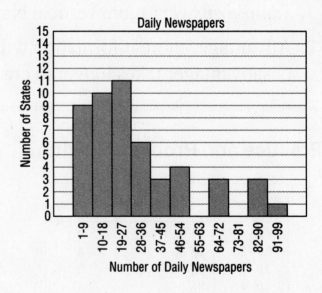

(continued)

6.

Interval	Democrat Tally	Democrat Frequency	Republican Tally	Republican Frequency
0 - 2	卌 卌 卌 卌 卌 l	26	卌 卌 卌 卌 l	21
3 - 5	卌 卌 ll	12	卌 卌 卌	15
6 - 8	卌 l	6	卌 ll	7
9 - 11	lll	3	ll	2
12 - 14		0	lll	3
15 - 17	l	1	l	1
18 - 20	l	1		0
21 - 23		0		0
24 - 26	l	1	l	1

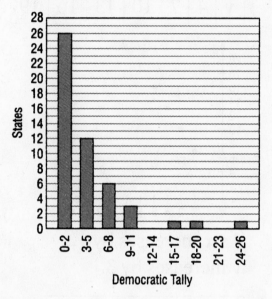

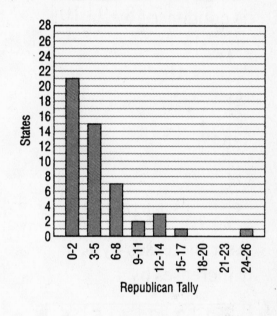

7. B

8. B

9.

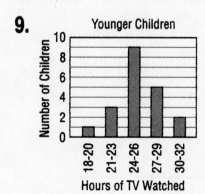

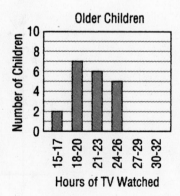

Answers for Spiral Review, page 230

1. $A = 0, B = 9, C = 1, D = 8, E = 2, F = 7,$
$G = 3, H = 6, I = 4, J = 5$

2.–5. *Sample answers*

2. $\frac{12}{16}$ **4.** $\frac{32}{60}$

3. $\frac{20}{28}$ **5.** $\frac{68}{76}$

6.

x	1	2	3	4	5	6
y	5	6	7	8	9	10

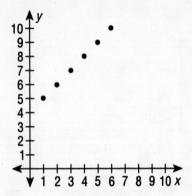

As x increases by 1,
y increases by 1.

8.

x	1	2	3	4	5	6
y	4	7	10	13	16	19

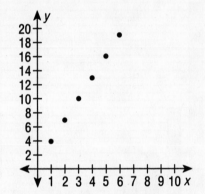

As x increases by 1,
y increases by 3.

7.

x	1	2	3	4	5	6
y	$\frac{1}{5}$	$\frac{2}{5}$	$\frac{3}{5}$	$\frac{4}{5}$	1	$1\frac{1}{5}$

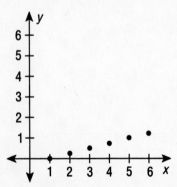

As x increases by 1,
y increases by $\frac{1}{5}$.

9.

x	1	2	3	4	5	6
y	10	8	6	4	2	0

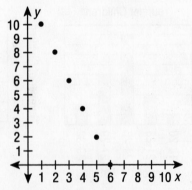

As x increases by 1,
y decreases by 2.

Answers for Lesson 5.3, pages 235–237

Ongoing Assessment

1. Prices should fall between the prices for compact cars and luxury cars.

2. The values should lie between the values for the compact and luxury cars.

Practice and Problem Solving

1. First

2. Median

3. Upper

4. Increasing

5.

6. The households have the same median, but the amount spent by household 1 varies more than the amount spent by household 2.

7. B

8. C

9. D

10. A

11.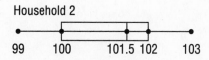

12.

13. *Sample answer:* the men's winning times were less than the women's; the women's winning times had a wider range than the men's.

14. D

15. C

16. Answers vary.

Answers for Lesson 5.4, pages 239–242

Practice and Problem Solving

1. $(-5, 1)$

2. $(3, 6)$

3. $(0, 0)$

4. $(7, -4)$

5. $(6, 0)$

6. $(-2, -2)$

7. II

8. III

9. I

10. IV

11. At least one is zero.

12.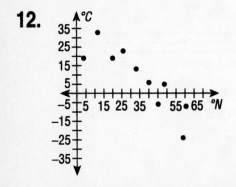

In general, as the latitude increases, the temperature decreases.

13.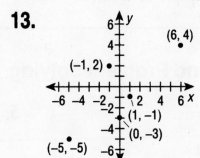

14.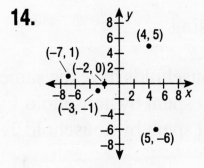

15.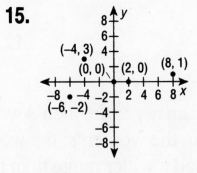

(continued)

Answers for Lesson 5.4, pages 239–242 (cont.)

16.

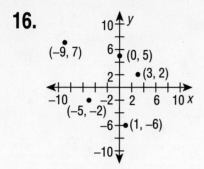

17.

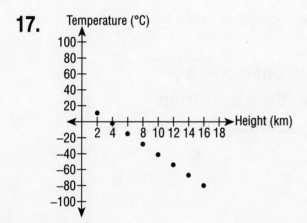

In general, as the height increases, the temperature decreases.

18.

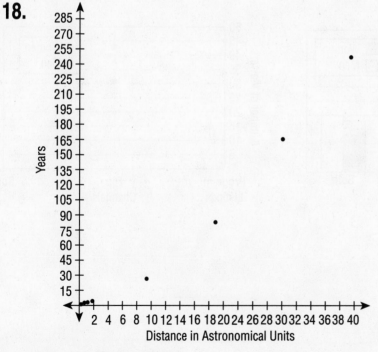

About 125 years

(continued)

Answers for Lesson 5.4, pages 239–242 (cont.)

19.

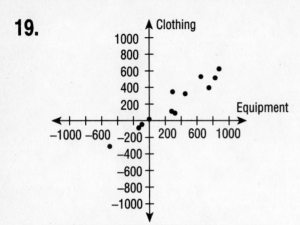

20. In general, as the profits from equipment increase, the profits from clothing increase; as the losses from equipment increase, the losses from clothing increase.

21. Between $50 and $250

22. A loss between $200 and $400

23. C

24. C

25. a.

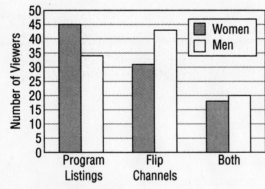 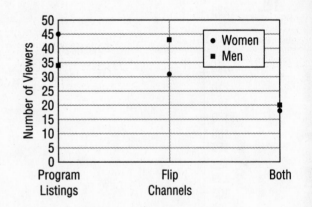

b. Answers vary.

Answers for Spiral Review, page 242

1. 7

2. 98

3. 10

4. 24

5. 50

6. 26

7. 768 1-ft-by-1-ft tiles,
192 2-ft-by-2-ft tiles,
48 4-ft-by-4-ft tiles,
12 8-ft-by-8-ft tiles

8. *D*

9. *E*

10. *G*

11. −9

12. −21

13. 0

14. 14

15. 14.4

16. $1\frac{1}{10}$

17. *Sample answer:*

Interval	Tally	Frequency
40 - 44	II	2
45 - 49	ЖГ I	6
50 - 54	ЖГ ЖГ II	12
55 - 59	ЖГ ЖГ II	12
60 - 64	ЖГ II	7
65 - 69	III	3

18. *Sample answer:*

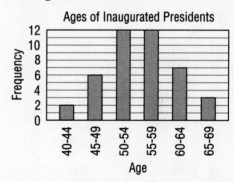

1. 3.6, 4, 4

2. $49, $45, $45

3. 17, 19, 21

4. *Sample answer:*

Interval	Tally	Frequency
0 - 9	IIII IIII IIII IIII	19
10 - 19	IIII IIII IIII	15
20 - 29	I	1
30 - 39	IIII III	8
40 - 49	IIII	5
50 - 59		0
60 - 69	I	1
70 - 79		0
80 - 89		0
90 - 99		0
100 - 109		0
110 - 119		0
120 - 129	I	1

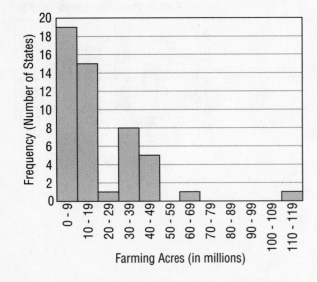

5.

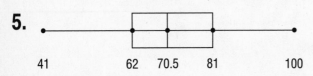

6. *Sample answer:* half of the soaps give between 62 and 81 handwashes per bar.

7.

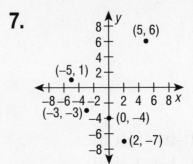

8.

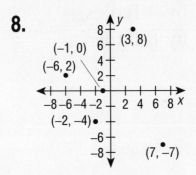

9.

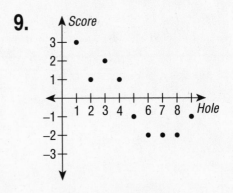

10. It appears that as you played, your scores tended to decrease.

Answers for Lesson 5.5, pages 245–247

Ongoing Assessment

1. About 13 pounds   **2.** About 12 pounds

3. About 30 quarts   **4.** About 50 quarts

Practice and Problem Solving

1.

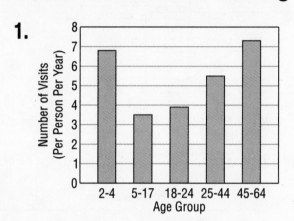

2. It would be confusing to graph the unequal age divisions along the horizontal axis.

3. In order: 5 ft/s, 10 ft/s, 9 ft/s, 6 ft/s, 2 ft/s

4. 4

5. 5 ft/s

6.

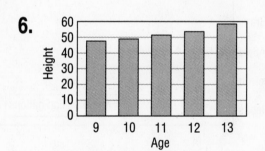 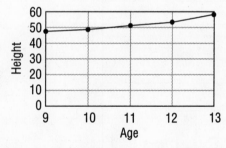

Height increases with each increase in age.

7.

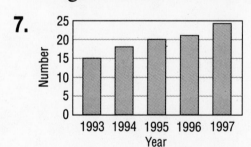 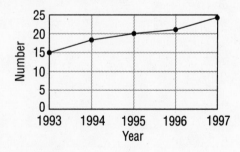

Number increases with each increase in year.

(continued)

8.

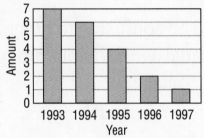

Value decreases with each increase in age.

9.

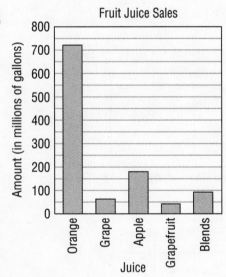

Amount decreases with each increase in year.

10.

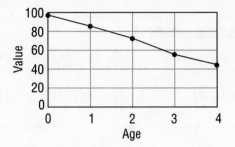

11.

12. Answers vary.

(continued)

Answers for Lesson 5.5, pages 245–247 (cont.)

13.

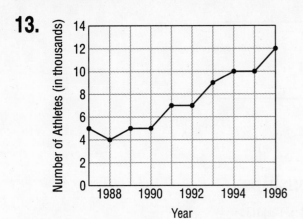

14. *Sample answer:*

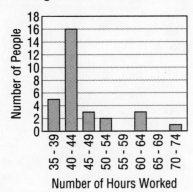

15.

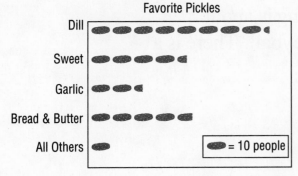

16. C

17. B

18. *Sample answer:*

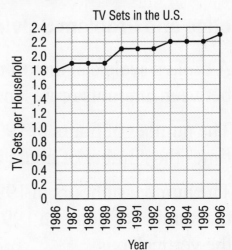

Answers for Lesson 5.6, pages 249–252

Practice and Problem Solving

1. A broken scale can distort the relationships between amounts represented by the graph.

2. The number of hours in 1996 appears to be three times the number of hours in 1991.

3. The number of hours in 1996 is about 1.5 times the number of hours in 1991. There is a break on the vertical axis.

4. The number of people who go bicycle riding appears to be $3\frac{1}{2}$ times the number of people who play volleyball.

5. The number who go bicycle riding is about 2.2 times the number who play volleyball. There is a break on the horizontal axis.

6.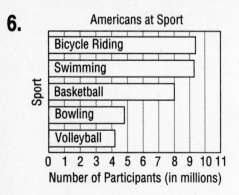

(continued)

Answers for Lesson 5.6, pages 249–252 (cont.)

7. It appears that people in the West prefer the Winter Olympics. The broken scale distorts the relationships between amounts represented by the graph.

8.

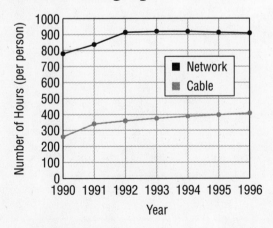

Olympics Preference

9. Spending increases.

10. $20 billion

11. $3\frac{1}{3}$ billion

12. The bottom graph

13. Answers vary.

14. B

15. D

16. C

17. a. Correct line graph:

b. Answers vary.

Answers for Spiral Review, page 252

1. Dining room, family/living room, kitchen, bedroom

2. $5\frac{1}{3}$ lb, $\frac{2}{3}$ lb

3. D

4. B

5. A

6. C

7.

c	1	2	3	4	5
Value	-3	-1	1	3	5

Expression: $2c - 5$

8. 8, 10, 10, 11, 11, 11, 11, 12, 12, 13, 13, 14, 14;
Mean: ≈ 11.54
Median: 11
Mode: 11

9. 0, 1, 2, 2, 3, 3, 4, 4, 4, 4, 4, 4, 5, 5, 5, 7;
Mean: 3.5625
Median: 4
Mode: 4

Answers for Communicating About Mathematics, page 253

1. There were 20 more men watching television in prime time.

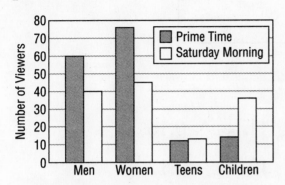

2. There were 46 more men than children watching during prime time. The difference would be smaller because there would be fewer men.

3. The average viewer watched 0.89 hours per week more in 1995 than in 1985.

4. The popularity of the drama series decreased from 1985 to 1995.

5.

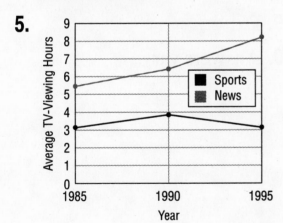

Answers for Lesson 5.7, pages 257–259

Ongoing Assessment

1. Answers vary.

2. Answers vary.

Practice and Problem Solving

1. A or B

2. D, E, or F

3. F

4. A

5. C

6. 3; answers vary.

7. B

8. C

9. A

10. $\frac{4}{77} \approx 0.05$

11. $\frac{12}{77} \approx 0.16$

12. $\frac{9}{77} \approx 0.12$

13. $\frac{2}{13} \approx 0.15$

14. $\frac{1}{9} \approx 0.11$

15. 0.47

16. 0.05

17. 0.95

18. C

19. D

20. Answers vary.

Answers for Chapter Review, pages 261–263

1. $350

2. $360

3. $400

4.

Interval	Tally	Frequency
0 - 3		0
4 - 7	II	2
8 - 11	II	2
12 - 15	IIII	5
16 - 19	IIII IIII	9
20 - 23	II	2

Answers vary.

5.

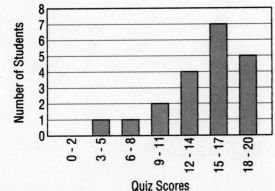
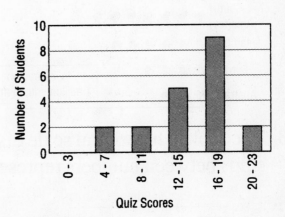

The second histogram shows fewer scores in the highest interval.

6. 34

7. 26

8. 40

9.

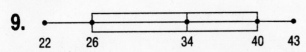

10. $(-22, -3), (-13, -25), (23, -5), (32, 0), (50, 10)$

(continued)

Answers for Chapter Review, pages 261–263 (cont.)

11.

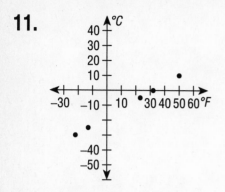

12. *Sample answer:*

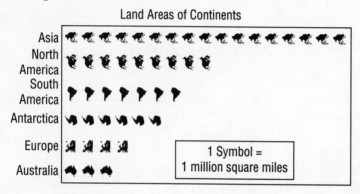

13. The broken horizontal scale distorts the relationships between numbers represented by the graph.

14.

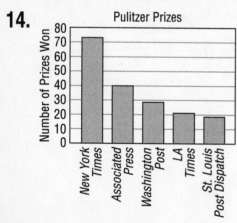

15. The scale on the graph is not broken.

16. 0.5; *left* and *right* are equally likely.

17. 1; certain to occur

Answers for Chapter Assessment, page 264

1. 407.1, 416, 452

2. *Sample answer:*

Interval	Tally	Frequency
0 - 4.9		0
5.0 - 9.9	I	1
10.0 - 14.9	JHT III	8
15.0 - 19.9	JHT	5
20.0 - 24.9		0
25.0 - 29.9	I	1

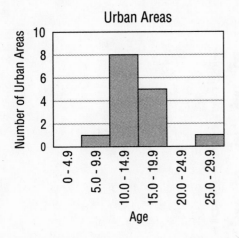

3.

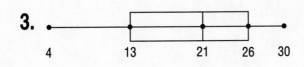

4.

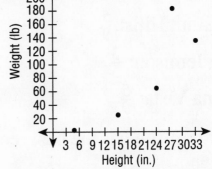

Taller dogs tend to be heavier.

5. *Sample answer:*

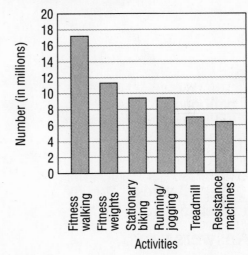

6. 0.08

7. 0.94

Answers for Standardized Test Practice, page 265

1. A

2. C

3. C

4. C

5. A

6. B

7. D

Answers for Lesson 6.1, pages 273–275

Ongoing Assessment

1. Chestnut Hills: $\frac{5}{9}$

   Mae Jemison: $\frac{3}{5}$

   Buena Vista: $\frac{5}{9}$

2. Mae Jemison has the best record because $\frac{3}{5} > \frac{5}{9}$.

Practice and Problem Solving

1. $\frac{1}{3}$

2. $\frac{1}{4}$

3. $\frac{5 \text{ balloons}}{3 \text{ balloons}}$

4. $\frac{1 \text{ in.}}{12 \text{ in.}}$

5. $\frac{9 \text{ feet}}{15 \text{ feet}}, \frac{3 \text{ yards}}{5 \text{ yards}}$; yes.

6. Answers vary.

7. No, units are not the same.

8. No, units are not the same.

9. Yes, units are the same.

10. No, units are not the same.

11. $\frac{3}{2}$

12. $\frac{4}{1}$

13. $\frac{1}{2}$

14. $\frac{2}{3}$

15. $\frac{2}{3}$

16. $\frac{1}{2}$

17. $\frac{6}{7}$

18. $\frac{11}{300}$

19. $\frac{5}{176}$

20. $\frac{108}{389}$

21. $\frac{21}{13}$

22. C

23. A

24. Yes. The ratio of the larger width to the smaller width is $\frac{80}{10}$, while the ratio of the larger length to the smaller length is $\frac{80}{11}$; $\frac{80}{10} \neq \frac{80}{11}$.

Answers for Lesson 6.2, pages 277–279

Ongoing Assessment

The 8-pound package because it is the least expensive per pound.

Practice and Problem Solving

1. A ratio is a fraction in which the numerator and denominator have the same unit of measure, a rate is a fraction in which the numerator and denominator have different units of measure. Answers vary.

2. **A.** $\dfrac{0.8 \text{ min}}{1 \text{ question}}$

   B. $\dfrac{1.25 \text{ questions}}{1 \text{ min}}$

   Answers vary.

3. 50 mi/h

4. $\dfrac{1}{3}$, ratio

5. $\dfrac{28 \text{ mi}}{1 \text{ gal}}$, rate

6. $\dfrac{72 \text{ beats}}{1 \text{ min}}$, rate

7. $\dfrac{33}{25}$, ratio

8. $\dfrac{3.5 \text{ inches}}{1 \text{ hour}}$

9. $\dfrac{\$.03}{1 \text{ sheet}}$

10. $\dfrac{650 \text{ mi}}{1 \text{ h}}$

11. $\dfrac{\$15}{1 \text{ day}}$

12. B; it has a lower rate: $.210 per oz compared to $.218 per oz.

13. C; it has the lowest rate: $.395 per lb compared to $.54 per lb for A and $.42 per lb for B.

14. B; it has a lower rate: $4.00 per gal compared to $4.50 per gal.

15. 2500 calories per day; 10,000 calories per day

(continued)

16. 16.547 mi/h

17. 46.4 hours per year, about 32 movies per year

18. 29 students per teacher

19. Shana. She types more words per minute (48) than Chung (42) and Randy (45).

20. B

21. C

22. $\dfrac{1 \text{ picture}}{0.001 \text{ second}}$; $\dfrac{1 \text{ picture}}{0.001 \text{ second}}$; the slower the shutter speed, the more likely the picture is to be blurred.

Answers for Spiral Review, page 280

1. 81

2. 1

3. 256

4. 8

5. 0.56

6. 0.58

7. 0.27

8. 0.64

9. $\frac{9}{10}$

10. -20

11. 25

12. 15

13. -10

14.

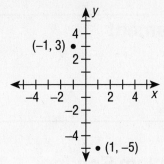

15.

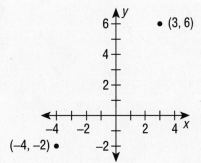

16.

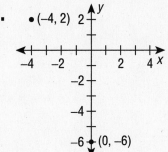

Answers for Lesson 6.3, pages 285–287

Practice and Problem Solving

1. B

2. C

3. A

4. B, 4

5. C, 1

6. A, 9

7. False

8. False

9. True

10. $\frac{n}{9} = \frac{10}{18}, 5$

11. $\frac{p}{21} = \frac{11}{33}, 7$

12. $\frac{35}{25} = \frac{z}{5}, 7$

13. $\frac{9}{6} = \frac{m}{18}, 27$

14. $\frac{4}{x} = \frac{6}{24}, 16$

15. $\frac{28}{w} = \frac{36}{9}, 7$

16. $\frac{27}{54}$

17. $\frac{13}{52}$

18. $\frac{19}{57}$

19. 1

20. 12

21. 1

22. 2

23. 12

24. 72

25. Answers vary.

26. 0.11

27. 55

28. 0.22

29. 5.5 in.

30. C, 24

31. D

32. 7200 feet

Answers for Lesson 6.4, pages 289–292

Ongoing Assessment

1. Should write $\frac{4 \times 2}{5 \times 3}$, not $\frac{4 \times 3}{5 \times 2}$.

2. When multiplying fractions, multiply numerators and multiply denominators.

Practice and Problem Solving

1. **a.** $\frac{16}{52} = \frac{n}{52}$, 16

   b. $13n = 208$, 16

2. $\frac{3\,\text{ft}}{7\,\text{ft}} = \frac{3}{7}$, $\frac{6\,\text{ft}}{14\,\text{ft}} = \frac{3}{7}$

3. D

4. B

5. C

6. A

7. True

8. False

9. False

10. True

11. True

12. False

13. False

14. True

15. 12

16. 9

17. 87

18. 32

19. 4

20. 7

21. 4.5

22. 9.3

23. $\frac{54\,\text{in.}}{90\,\text{in.}} \overset{?}{=} \frac{57\,\text{in.}}{95\,\text{in.}}$
    $\frac{3}{5} = \frac{3}{5}$

24. $\frac{8\,\text{m}}{3\,\text{m}} \overset{?}{=} \frac{16\,\text{m}}{6\,\text{m}}$
    $\frac{8}{3} = \frac{8}{3}$

25. 24.14

26. 3.94

27. 63.64

28. 22.12

29. True, answers vary.

30. False, answers vary.

(continued)

Answers for Lesson 6.4, pages 289–292 (cont.)

31. 9.6 h

32. a. 9; Number that prefer 80's
   b. 4; Number that prefer 90's
   c. 3; Number that prefer 60's
   d. 7; Number that prefer 70's

33. C

34. B

35. C

36. about 1.6, about 8 in.

37. $a = 5, b = 4, c = 25, d = 20, e = 6,$
   $f = 30, g = 3, h = 15$

1. $\frac{17}{24}$

2. $\frac{31}{45}$

3. $\frac{17}{30}$

4. $\frac{43}{60}$

5. $5\frac{1}{3}$

6. -5

7. -15

8. -9

9. 5

10. **Verbal Model:** $\boxed{\text{Amount already saved}} + \boxed{\text{Amount left to save}} = \boxed{\text{Cost of bicycle}}$

   Labels: Amount already saved $= \$95$
   Amount left to save $= x$
   Cost of bicycle $= \$189$

   Equation: $95 + x = 189$

   Solution: $\$94$

11. $\frac{1}{2}$

12. $\frac{1}{3}$

13. $\frac{5}{16}$

Answers for Mid-Chapter Assessment, page 293

1. $\frac{49}{300}$

2. $\frac{32}{25}$

3. $\frac{11}{15}$

4. $\frac{74 \text{ cal}}{1 \text{ oz}}$

5. $\frac{500 \text{ pieces of paper}}{1 \text{ in.}}$

6. $\frac{k}{12} = \frac{5}{60}$, 1

7. $\frac{x}{6} = \frac{6}{2}$, 18

8. $\frac{4}{7} = \frac{m}{35}$, 20

9. $\frac{72}{9} = \frac{s}{4}$, 32

10. True

11. False

12. True

13. Sam Houston Tollway: $\frac{\$.19}{1 \text{ mile}}$

    Delaware Turnpike: $\frac{\$.11}{1 \text{ mile}}$

    Daniel Boone Parkway: $\frac{\$.03}{1 \text{ mile}}$

    Chicago Skyway: $\frac{\$.26}{1 \text{ mile}}$

    Judging by dollars per mile, Chicago Skyway is the most expensive.

14. b; 153 steps

15. 857,143 lb

Answers for Lesson 6.5, pages 295–297

Ongoing Assessment
1. 50,500 years
2. Answers vary.

Practice and Problem Solving

1. 330 min or $5\frac{1}{2}$ hr
2. $2928
3. 102 inches
4. 24 cm
5. C, 9 in.
6. B, 225 ft^2
7. Answers vary.

8. $0.08
9. 625 ft^2
10. 286 shillings
11. 72.75 people per square mile
12. B
13. C
14. 6 ounces

Answers for Lesson 6.6, pages 301–303

Practice and Problem Solving

1. Answers vary.

2. Yes

3. No

4. $\overline{LM}$ corresponds to $\overline{DE}$, $\overline{MN}$ corresponds to $\overline{EF}$, $\overline{LN}$ corresponds to $\overline{DF}$.

5. 53°

6. 32

7. $\frac{24}{32} = \frac{15}{20} = \frac{3}{4}$

8. $\angle A$ and $\angle X$ are 40°, $\angle B$ and $\angle Y$ are 100°; $\angle C$ and $\angle Z$ are 40°.

$$\frac{AB}{XY} \overset{?}{=} \frac{BC}{YZ} \overset{?}{=} \frac{AC}{XZ}$$

$$\frac{38}{57} \overset{?}{=} \frac{38}{57} \overset{?}{=} \frac{58}{87}$$

$$\frac{2}{3} = \frac{2}{3} = \frac{2}{3}$$

Corresponding angles have the same measure and corresponding sides are proportional, so the triangles are similar.

(continued)

Answers for Lesson 6.6, pages 301–303 (cont.)

9. All angles are 90°.

$$\frac{GM}{UP} \overset{?}{=} \frac{ML}{TU} \overset{?}{=} \frac{KL}{ST} \overset{?}{=} \frac{KJ}{RS} \overset{?}{=} \frac{HJ}{QR} \overset{?}{=} \frac{GH}{PQ}$$

$$\frac{39}{52} \overset{?}{=} \frac{36}{48} \overset{?}{=} \frac{24}{32} \overset{?}{=} \frac{45}{60} \overset{?}{=} \frac{15}{20} \overset{?}{=} \frac{81}{108}$$

$$\frac{3}{4} = \frac{3}{4} = \frac{3}{4} = \frac{3}{4} = \frac{3}{4} = \frac{3}{4}$$

Corresponding angles have the same measure and corresponding sides are proportional, so the polygons are similar.

10. No, corresponding angles may not have the same measure.

11. Yes, all angles are 90° and corresponding sides are proportional.

12. No, corresponding sides may not be proportional.

13. 27

14. $\angle P$ is 40°, $\angle U$ is 25°, $PQ = 10$, $SU = 42$

15. $\angle F$ is 45°, $\angle G$ is 45°, $FG = 21$, $XZ = 25$

16. $\angle Y$ is 45°, $\angle H$ is 135°, $FG = 56$, $GH = 49$, $WZ = 30$

17. $344\frac{1}{2}$ ft

18. B

19. A

20. 16 ft

Answers for Spiral Review, page 304

1. 2-0.5 gal and 3-2 gal containers

2. <

3. =

4. >

5. B

6. C

7. A

8. D

9. $\frac{5}{9}$

10. $\frac{11}{7}$

11. $\frac{1}{7}$

12. $8 + n = -5, -13$

13. $-9 = n - 16, 7$

14.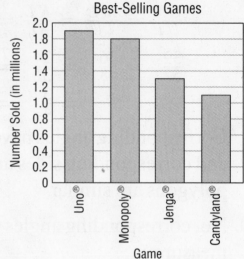

15. $\frac{7}{3}$

16. 4

17. $\frac{1}{12}$

Answers for Communicating About Mathematics, page 305

1. $\frac{32}{7}$; the units are the same.

2. Answers vary.

3. $\dfrac{2\frac{1}{2} \text{ trips}}{1 \text{ battery charge}}$; rate because the units are different.

4. Rate; the units are different.

5. About 3 minutes; answers vary.

6. 600,000; $\frac{1}{10}$ of all cars produced will be electrically powered.

Answers for Lesson 6.7, pages 307–309

Ongoing Assessment

1. 1.95 in.

2. 91 miles

Practice and Problem Solving

1. C

2. A

3. B

4. 2.5 cm

5. 40 in.

6. 56 ft

7. Answers vary.

8. 12 mi

9. $2\frac{2}{3}$ mi

10. 1.6 mi

11. 21.6 mi

12. 18 ft

13. 384 ft

14. 80 ft

15. 299 mi

16. 260 mi

17. 156 mi

18. 325 mi

19. 1.5 in.

20. 80 ft

21. 100 ft

22. 140 ft

23. D

24. Answers vary.

1. $\frac{10}{3}$

2. $\frac{360}{17}$

3. $\frac{64}{3}$

4. $\frac{5}{4}$

5. $\frac{1}{4}$

6. $\frac{84 \text{ beats}}{1 \text{ minute}}$

7. $\frac{26 \text{ miles}}{1 \text{ gallon}}$

8. a; the 32-oz can

9. 3

10. $\frac{2}{5}$

11. $\frac{7}{9}$

12. 33

13. 65

14. $\frac{7}{4}$ or $1\frac{3}{4}$

15. 51

16. 10

17. **Verbal Model:** $\dfrac{\boxed{\text{Gallons of water (1)}}}{\boxed{\text{Gallons of water (2)}}} = \dfrac{\boxed{\text{Flow time in seconds (1)}}}{\boxed{\text{Flow time in seconds (2)}}}$

   Labels: Gallons of water (1) = 750,000

   Gallons of water (2) = 3,000,000

   Flow time in seconds (1) = 1

   Flow times in seconds (2) = x

   Equation: $\dfrac{750,000}{3,000,000} = \dfrac{1}{x}$

   Solution: 4 seconds

18. $\dfrac{22}{33} \overset{?}{=} \dfrac{44}{66} \overset{?}{=} \dfrac{38}{57}$

   $\dfrac{2}{3} = \dfrac{2}{3} = \dfrac{2}{3}$

   Corresponding angles have the same measure and corresponding sides are proportional, so the triangles are similar.

19. 4 in. long and 3 in. wide

Answers for Chapter Assessment, page 314

1. Rate, units are different. $\frac{2 \text{ cats}}{3 \text{ dogs}}$

2. Rate, units are different. $\frac{10 \text{ questions}}{9 \text{ minutes}}$

3. Ratio, units are the same. $\frac{3}{5}$

4. Rate, units are different. $\frac{11 \text{ dollars}}{2 \text{ hours}}$

5. $5.50 per hour

6. B; unit rate is less.

7. B; unit rate is less.

8. A; unit rate is less.

9. 14

10. 91

11. 4

12. 4

13. 5.6

14. 21.2

15. $\angle A = 60°, \angle F = 30°, AC = 8, DE = 8$

16. $\angle H = 45°, \angle L = 135°, FJ = 15, FG = 12,$
    $NM = 45$

17. **Verbal Model:** $\quad \dfrac{10}{19} = \dfrac{\boxed{\text{Width of flag in feet}}}{\boxed{\text{Length of flag in feet}}}$

    Labels: Width of flag in feet $= x$
    Length of flag in feet $= 2$

    Equation: $\dfrac{10}{19} = \dfrac{x}{2}$

    Solution: $1\frac{1}{19}$ or ≈ 1.05 feet

18. $\frac{1}{48}$

19. 72 ft

20. $\frac{25}{24}$ in. or $1\frac{1}{24}$ in.

Answers for Standardized Test Practice, page 315

1. C

2. B

3. C

4. A

5. C

6. C

7. C

8. B

9. B

10. C

Answers for Cumulative Review, pages 316 and 317

1. Shape with 4 sides

2. Shape with 5 sides

3. Shape with 3 sides

4. △ is 2, ⬜ is 9, ▢ is 1, ○ is 3, ◸ is 6, ⬠ is 4, ▱ is 8

$$2\overline{)93} \quad \begin{array}{r} 46.5 \\ \underline{8} \\ 13 \\ \underline{12} \\ 1 \end{array}$$

5. Kay, May, Ray, and Fay

6. 12

7. 16

8. 78

9. $2^3 \cdot 3^2$

10. $3 \cdot 5^2$

11. $2^4 \cdot 7$

12. $3^3 \cdot 7$

13. 2, 42

14. 8, 168

15. 5, 150

16. C

17. B

18. D

19. A

20. $\frac{5}{9}$, 1.75, $\frac{9}{5}$, 2.65, $\frac{8}{3}$, $\frac{7}{2}$

21. 1.05, 1.1, $\frac{9}{8}$, $\frac{6}{5}$, 1.25, $\frac{3}{2}$

22. $\frac{5}{6}$

23. $\frac{3}{20}$

24. $1\frac{7}{8}$

25. $10\frac{1}{24}$

26. $\frac{3}{14}$

27. $8\frac{1}{2}$

28. $1\frac{7}{9}$

29. $1\frac{1}{24}$

30.

x	1	2	3	4	5	6
	−4	−1	2	5	8	11

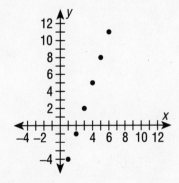

(continued)

Answers for Cumulative Review, pages 316 and 317 (cont.)

31.

x	1	2	3	4	5	6
	4	2	0	-2	-4	-6

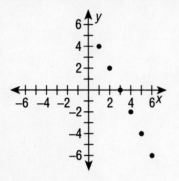

36.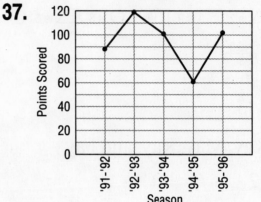

32. -160 ft

33. -25 ft

34. 25, 25, 25

35. 136, 133, 131

37.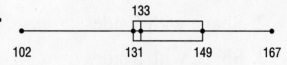

38. 1.09, 1.42, 1.20, 1.27, 1.24

39. 5

40. 72

41. 12

Answers for Lesson 7.1, pages 323–325

Ongoing Assessment

1. $\frac{1}{100}$

2. $\frac{1}{4}$

3. $\frac{1}{3}$

4. $\frac{2}{5}$

5. $\frac{1}{2}$

6. $\frac{3}{4}$

Practice and Problem Solving

1. *Sample answer:* Percent means per hundred.

2. Meaning of 36%, divide numerator and denominator by 4.

3. Meaning of 0.625, divide numerator and denominator by 10, meaning of 62.5%.

4. $0.55, \frac{11}{20}$

5. Answers vary.

6. 84%

7. $0.08, \frac{2}{25}$

8. $0.22, \frac{11}{50}$

9. $0.45, \frac{9}{20}$

10. $0.9, \frac{9}{10}$

11. $\frac{78}{100}, 78\%$

12. $\frac{4}{100}, 4\%$

13. $\frac{9.5}{100}, 9.5\%$

14. $\frac{40}{100}, 40\%$

15. 56%

16. 6%

17. 87.5%

18. 75%

19. 44%

20. 55%

21. 27%

22. $33\frac{1}{3}\%$

23. $\frac{14}{25}, 68\% = 0.68 = \frac{68}{100} = \frac{34}{50} \neq \frac{14}{25}$

24. $9.1\%, \frac{20}{22} = \frac{10}{11} \approx 0.909 = 90.9\% \neq 9.1\%$

(continued)

25.–28. Answers vary.

29. Geography: $\frac{25}{160} = 15.625\%$

Sports & Leisure: $\frac{25}{160} = 15.625\%$

Science & Nature: $\frac{60}{160} = 37.5\%$

History: $\frac{30}{160} = 18.75\%$

Entertainment: $\frac{10}{160} = 6.25\%$

Arts & Literature: $\frac{10}{160} = 6.25\%$

100%; everyone responded to the survey.

30. The Science & Nature percent is twice as large as the History percent, which means that twice as many people chose Science & Nature as chose History.

31. B

32. A

33. 15.8%, 15.8%, 10.5%, 26.3%, 6.6%

Answers for Lesson 7.2, pages 329–331

Ongoing Assessment

1. Answers vary.

2. Answers vary.

Practice and Problem Solving

1. 2.5 mi, 75 mi

2. 0.8 sandwich, 12 sandwiches

3. 0.45 orange, 27 oranges

4. 2 baseball cards

5. a. 66 baseball cards

    b. $0.33 \times 200 = 66$

6. Answers vary.

7. 54%, 27 hours

8. 28%, 42 kg

9. 92%, 460 cups

10. 40 cars

11. 40 questions

12. $.24

13. 0.5, 20

14. 2.5, 100

15. 8, 320

16. 68

17. 156

18. 12

19. 120

20. 24

21. 45

22. 24

23. 54

24. 60.5

25. 34

26. 316.22

27. 219

28. 517.5 metric tons, 315 metric tons, 135 metric tons

29. D

30. B

31. Cat: 6 pounds; human baby: 8.64 pounds; the fluid in the human baby is about one and one-half times the fluid in the cat.

Answers for Spiral Review, page 332

1.–4. *Sample answers*

1. $\dfrac{9}{36}, \dfrac{1}{4}$

2. $\dfrac{10}{15}, \dfrac{2}{3}$

3. $\dfrac{16}{28}, \dfrac{4}{7}$

4. $\dfrac{25}{45}, \dfrac{5}{9}$

5. $1\dfrac{1}{6}$ feet

6. *Sample answer:*

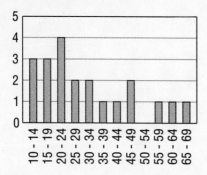

7. 35

8. 18

9. 27

10. 4.6

Answers for Lesson 7.3, pages 337–339

Ongoing Assessment

1.
$$
\begin{aligned}
200\% \text{ of } 48 &= 96 \\
+\ 20\% \text{ of } 48 &= \ \ 9.6 \\
\hline
220\% \text{ of } 48 &= 105.6
\end{aligned}
$$

2. $\frac{1}{3}\%$ is one third of 1%.

1% of $60 = 0.6$

One third of 0.6 is 0.2.

So, $\frac{1}{3}\%$ of 60 is 0.2.

3. 324% is about 325%.
$$
\begin{aligned}
300\% \text{ of } 40 &= 120 \\
+\ 25\% \text{ of } 40 &= \ \ 10 \\
\hline
325\% \text{ of } 40 &= 130
\end{aligned}
$$

So, 324% of 40 is about 130.

Practice and Problem Solving

1. 164%

2. $\frac{1}{3}\%$

3. B

4. D

5. A

6. C

7. 390

8. 1

9. 72 inches

10. Answers vary.

11. a. 1.7
  b. $\frac{17}{10}$

12. a. 0.008
  b. $\frac{1}{125}$

13. a. 0.00375
  b. $\frac{3}{800}$

14. a. 2.32
  b. $\frac{58}{25}$

15. a. 4.5
  b. $\frac{9}{2}$

(continued)

Answers for Lesson 7.3, pages 337–339 (cont.)

16. 195%

17. 650%

18. $\frac{3}{4}$%

19. 0.875%

20. 324%

21. 55 points

22. $720

23. $\frac{1}{2}$ day

24. >

25. >

26. <

27. =

28. 0.25, 0.5, 0.75, 1. To get each number after the first, add 0.25 to the preceding number; 1.25, 1.50.

29. 165, 240, 325, 420. To get each number after the first, add 75, add 85, add 95, and so on, to the preceding number; 525, 640.

30. New York: about 16%
Chicago: about 4%
Boston: about 2.5%
Houston: about 1.9%
Atlanta: about 1.4%

31. About 13,200

32. $21

33. D

34. C

35. 16.1 hands. 64.4 in.; answers vary.

Answers for Lesson 7.4, pages 343–346

Ongoing Assessment

1. Answers vary.

2. Answers vary.

Practice and Problem Solving

1. 17, 20; 85; 85%

2. 300, 33; 99; 99

3. 25, 20; 125; 125

4. Cats: 66,054,570
Dogs: 32.9%
Small animals: 12,750,480
Parakeets: 22.6%

5. $\frac{36}{150} = \frac{p}{100}$, 24%

6. $\frac{4}{b} = \frac{8}{100}$, 50

7. 750

8. 7.2

9. 200

10. $24.50

11. 18

12. 20%

13. 25%

14. 350

15. 9

16. 12

17. 8, 8, 8; each time the percent doubles and the base is halved, the result remains the same.

18. 1, 4, 9; each time the percent increases by 1 and the base increases by 100, the result is the next successive perfect square.

19. 15, 30, 45; each time the percent increases by 25, the result increases by 15.

20. 219

21. 72

(continued)

Answers for Lesson 7.4, pages 343–346 (cont.)

22. China: about 20.9%
India: about 16.7%
U.S.: about 4.7%
Indonesia: about 3.4%
Mexico: about 1.7%

23. A

24. C

25. 5%

26. 2.5%

Answers for Spiral Review, page 346

1. $\frac{2}{3}$

2. $\frac{11}{24}$

3. $1\frac{2}{3}$

4. $6\frac{1}{15}$

5. $\frac{25}{36}$ ft^2

6. $12.32\frac{8}{25}$ m^2

7. $3\frac{3}{8}$ yd^2

8. -24

9. -7

10. 16

11. -2

12. -13

13. 61

14.

15. about $\dfrac{84 \text{ perfect games}}{1 \text{ day}}$

16. 1 foot

Answers for Mid-Chapter Assessment, page 347

1. 12.4%

2. 35%

3. 6.7%

4. 75%

5. 289%

6. $\frac{5}{9}$%

7. 3.7%

8. 406.25%

9. 75%, 180 feet

10. 65%, 16.25 muffins

11. 22%, 35.2 pounds

12. <

13. >

14. =

15. 132

16. 600

17. 117

18. 8

19. $100

20. $24.75

21. **a.** 25,000

    b. 1900

Ongoing Assessment

1.

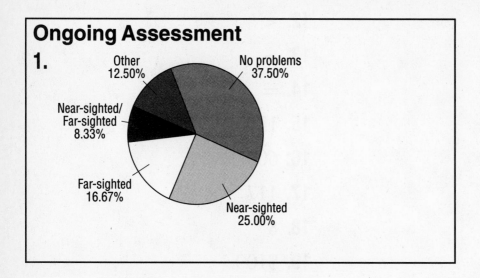

Other
12.50%

No problems
37.50%

Near-sighted/
Far-sighted
8.33%

Far-sighted
16.67%

Near-sighted
25.00%

Practice and Problem Solving

1. See page 348 of the textbook. See if the sum of the angles is 360°.

2. 144°, 35%

3. 36°, 198°, 72°, $\frac{1}{5}$, $\frac{3}{20}$

4. $\frac{1}{5}$, $\frac{1}{10}$, 162°, 36°

5. Yes; the sum of the numbers in the six categories equals the number of people surveyed.

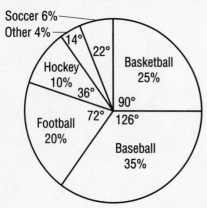

Soccer 6%
Other 4%
14°
22°
Hockey
10%
36°
Basketball
25%
90°
Football
20%
72°
126°
Baseball
35%

(continued)

Answers for Lesson 7.5, pages 349–352 (cont.)

6. No; the sum of the numbers in the four categories does not equal the number of people surveyed. This means a person could choose more than one category.

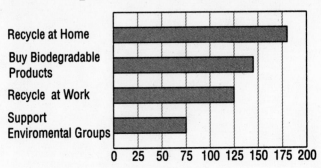

7. About 95

8. About 78

9. About 4

10. 57.6°

11.

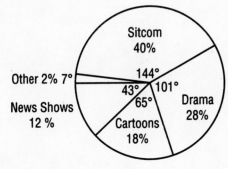

12. 40%

13. 40% + 28% + 18% + 12% + 2% = 100%

14.

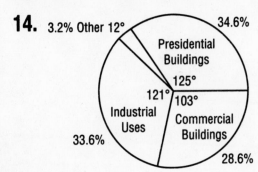

(continued)

Answers for Lesson 7.5, pages 349–352 (cont.)

15.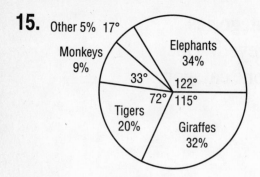

16. D

17. C

18. D

19. C

20.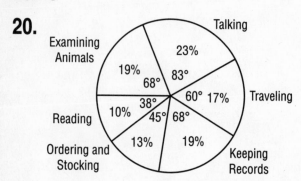

Answers for Spiral Review, page 352

1. 48; *Sample answer:* Each successive number is three less than the preceding number.

2. *Sample answer:* Letters of the alphabet; 1st, 25th, 3rd, 23rd, . . .

3. 100

4. 90

5. 320

6. 24

7. 2, 1, 0, −1, −2

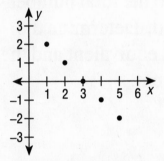

Beginning with 2, each succeeding number decreases by 1.

8. 1, 3, 5, 7, 9

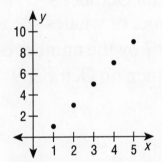

Beginning with 1, each succeeding number increases by 2.

9. 4.25, 4.3, 4.4

10. $2\frac{2}{3}$ hours or 2 hours and 40 minutes

11. 10 feet

12. $\frac{4}{5}$; $45\% = 0.45 = \frac{45}{100} = \frac{9}{20} \neq \frac{4}{5}$

13. 625%; $\frac{25}{40} = \frac{5}{8} = 0.625 = 62.5\% \neq 625\%$

Answers for Communicating About Mathematics, page 353

1. $\dfrac{29}{3.97} \approx 7.3$ sightings per hour;
You want a rate of $\dfrac{\text{Sightings}}{\text{Hour}}$.
Divide the number of sightings by the number of hours.

2. $\dfrac{4}{65} \approx 0.06$; You want a ratio of $\dfrac{\text{Whales on October 7}}{\text{Whales on October 9}}$. Divide the number of whales seen on October 7 by the number of whales seen on October 9.

3. $\dfrac{63 + 65}{204} = \dfrac{p}{100}$; 62.75%

4. 5.25 hours; You want a proportion that equates the number of whales per hour to the 65 whales observed in x hours.

5. 4.35%; Start with a ratio of the number observed diving to the total number observed. Determine the decimal equivalent and then the percent.

Answers for Lesson 7.6, pages 355–357

Ongoing Assessment

1.

Month	Beginning Balance	Interest	Ending Balance
5	$306.05	$1.53	$307.58
6	$307.58	$1.54	$309.12
7	$309.12	$1.55	$310.67

The interest increases $.01 every month, except every 4th month when it is the same as the previous month.

2. $1.55

Practice and Problem Solving

1. C

2. A

3. D

4. B

5. No, $30 in interest is 60% of $50. Correct answer: $2.50.

6. No, $150 in interest is 150% of $100. Correct answer: $102.50.

7. $\frac{1}{4}$

8. $\frac{1}{2}$

9. $\frac{2}{3}$

10. $\frac{3}{4}$

11. $20

12. $80

13. $200

14. $240

15. $320

16. $440

17. $560

18. $720

19. About $35,333,000,000

20. $10, $610

21. $18.75, $1268.75

(continued)

Answers for Lesson 7.6, pages 355–357 (cont.)

22. $577.50, $7577.50

23. $1350; $13,350

24. $168.75, $4668.75

25. $135, $1035

26. 2.5%

27. $12.89

28. $12.60

29. $13.08

30. $14.00

31. A

32. C

33. b, c, a, d

Answers for Lesson 7.7, pages 359–361

Ongoing Assessment

1. 1985 to 1990

2. 1985 to 1990

Practice and Problem Solving

1. C

2. B

3. A

4. D

5. True; 100% of a number $= 1\times$ the number.

6. False; "original" should replace "new."

7. A student who has less than 100 points should choose b, a student who has more than 100 points should choose a.

8. 75% increase

9. 50% increase

10. 75% increase

11. 60% decrease

12. 75% decrease

13. 25% increase

14. 12% decrease

15. 9.1% decrease

16. 5.8% increase

17. About 50%

18. About 30%

19. About 75%

20. About 10%

21. About 10.0%

22. About 2.1%

23. About 12.2%

24. C

25. D

26. 1986 to 1990: 516.9% increase
    1990 to 1994: 54.4% decrease

Answers for Chapter Review, pages 363–365

1. 32%

2. 0.68, 68%

3. 0.4, $\frac{2}{5}$

4. 0.15, $\frac{3}{20}$

5. 0.72, $\frac{18}{25}$

6. 0.37, $\frac{37}{100}$

7. **a.** 28%

   b. 126 feet

8. 8, 17, 27, 38. To get each number after the first, add 9, add 10, add 11, and so on, to the preceding number. 50, 63

9. 5.88

10. 73.5

11. 0.025

12. $\frac{2}{5}$

13. 300

14. A little more than 120

15. 60%

16. 21.6

17. 180

18.

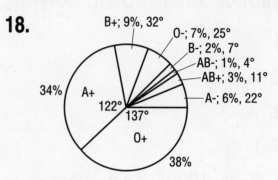

19. $2.25

20. $3120

21. 11.1%

22. 6.8%

23. 18.5%

Answers for Chapter Assessment, page 366

1. a. 0.74   **b.** $\frac{37}{50}$

2. a. 0.05   **b.** $\frac{1}{20}$

3. a. 0.006   **b.** $\frac{3}{500}$

4. a. 2.16   **b.** $\frac{54}{25}$

5. 88%

6. $43\frac{3}{4}\%$

7. 440%

8. 0.625%

9. $\frac{11}{40}\%$

10. 4%

11. 6.2%

12. 303%

13. 18 hours

14. $7.25

15. 11 goals

16. 1 person

17.

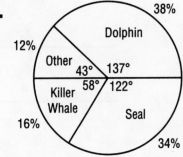

18. $44

19. 105

20. 550

21. 336%

22. 0.5%

23. 86.5%

24. 38.9%

25. 1993–1994, 30.0%

Answers for Standardized Test Practice, page 367

1. A

2. C

3. C

4. C

5. D

6. D

7. D

8. D

9. B

10. C

11. B

12. C

Answers for Lesson 8.1, pages 375–377

Ongoing Assessment

Sample answer:

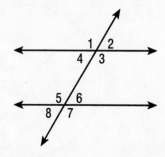

1. ∠1 and ∠3, ∠2 and ∠4, ∠5 and ∠7, ∠6 and ∠8

2. ∠1 and ∠2, ∠1 and ∠4, ∠2 and ∠3, ∠3 and ∠4, ∠5 and ∠6, ∠5 and ∠8, ∠6 and ∠7, ∠7 and ∠8

3. Angles 1, 3, 5, and 7 are congruent to each other. Angles 2, 4, 6, and 8 are congruent to each other.

Practice and Problem Solving

1. ∠1 and ∠3, ∠2 and ∠4, ∠5 and ∠7, ∠6 and ∠8

2. ∠1 and ∠2, ∠1 and ∠4, ∠2 and ∠3, ∠3 and ∠4, ∠5 and ∠6, ∠5 and ∠8, ∠6 and ∠7,∠7 and ∠8

3. **a.** False
   b. False
   c. True
   d. True

4. **a.** 125°
   b. 55°
   c. 125°

5. 60°

6. 90°

7. 90°

8. 60°

9. Never

10. Always

(continued)

Answers for Lesson 8.1, pages 375–377 (cont.)

11. Sometimes

12. Sometimes

13. Never

14. $\angle 1 = 40°$, $\angle 2 = 140°$

15. $\angle 1 = 45°$, $\angle 2 = 90°$

16. $\angle 1 = 90°$, $\angle 2 = 90°$

17. $50°$

18. $170°$

19. Answers vary.

20. West 6th Street

21. Oak St., Main St.

22. a: $45°$, b: $45°$, c: $135°$

23. B

24. C

25. Yes; answers vary.

Answers for Lesson 8.2, pages 381–383

Ongoing Assessment

1. Figure E is at (2, 6).
   Figure F is at (3, 8).
   Figure G is at (4, 10).
   Figure H is at (5, 12).

2. Figure E is a translation of Figure C.
   Figure F is a translation of Figure B.
   Figure G is a translation of Figure E.
   Figure H is a translation of Figure D.

Practice and Problem Solving

1. 5 units to the right and 6 units up

2. 1 unit to the left and 5 units down

3. 1 and 3, 2 and 4, 5 and 7, 6 and 8

4. Answers vary.

5. 4 units to the left and 3 units down

6. 6 units to the right and 8 units up

7. (2, 2), (8, 2), (8, −4)

8. (−7, −3), (−6, −7), (−1, −1), (−2, −9)

9. Answers vary.

10. Yes, because you could slide the original figure to the last figure with every point moving in the same direction and for the same distance.

11. A and F; 36 right and 1 down
    B and C; 7 right and 2 down

(continued)

Answers for Lesson 8.2, pages 381–383 (cont.)

12. C

13. B

14. a. From hole A, move 2 units to the left and 3 units up to drill hole B. From hole A, move 2 units to the right and 3 units down to drill hole C.

b. Answers vary.

Answers for Spiral Review, page 384

1. 1 quarter, 2 dimes, 4 nickels

2. 17

3. 6

4. 18

5. 42

6. 40

7. 60

8. $\frac{2}{5}$

9. $\frac{1}{21}$

10. $1\frac{11}{40}$

11. 2

12. 4

13. $1\frac{25}{27}$

Answers for Lesson 8.3, pages 387–389

> **Ongoing Assessment**
>
> **1.** The circumference of the ball is more than half of the circumference of the basketball hoop.
>
> **2.** The radius of a basketball is about 4.8 inches which is a little more than half of the radius of a basketball hoop.

Practice and Problem Solving

1. Circumference is the perimeter of a circle. The diameter is the distance across the circle through its center. The radius is the distance from the center of the circle to any point on the circle.

2. 18.84 in.

3. 12.56 ft

4. 25.12 m

5. 14.92 in.

6. 5 cm

7. 31.4 in.

8. 100.48 cm

9. 62.8 ft

10. 31

11. Circumference doubles; circumference doubles; answers vary.

12. A

13. C

14. B

15. 1.80 cm

16. 15 m

17. 0.16 in.

18. 100.48 ft

19. 37.68 ft

20. A

21. C

22. 188.4 ft

Answers for Lesson 8.4, pages 391–394

> **Ongoing Assessment**
>
> 1. Georgia is about 300 mi high and 200 mi across. The area is about $300 \times 200 = 60{,}000 \text{ mi}^2$. Actual area of Georgia is $58{,}977 \text{ mi}^2$.
>
> 2. *Sample answer:* WA, OR, WY, CO, NM, KS, ND, SD, MO, CT

Practice and Problem Solving

1. Base is 6 units. Height is 3 units.

2. 18 units^2

3. 18 units^2

4. 9 is the height, not 10; $A = 9 \cdot 15 = 135 \text{ in.}^2$

5. 12 cm   **6.** 70 m^2 **7.** 49 ft^2 **8.** 52.7 cm^2

9. *Sample answer:* $b = 4, h = 3$; $b = 2, h = 6$

10. *Sample answer:* $b = 10, h = 3$; $b = 6, h = 5$

11. *Sample answer:* $b = 12, h = 3$; $b = 18, h = 2$

12.

Base	Height	Area
5 units	1 units	5 units^2
5 units	2 units	10 units^2
5 units	4 units	20 units^2
5 units	8 units	40 units^2

Each time the height is doubled, the area doubles.

(continued)

Answers for Lesson 8.4, pages 391–394 (cont.)

13. 17 in.

14. 9.6 m

15. $4\frac{1}{8}$ ft

16. a. Triangles:
$$\tfrac{1}{2}(6)(7) = 21 \text{ units}^2;$$
Rectangle:
$$3(7) = 21 \text{ units}^2;$$
$$21 + 21 + 21 = 63 \text{ units}^2$$

  b. $A = b \cdot h$
$$= 9 \cdot 7 = 63 \text{ units}^2$$
The results are the same.

17. All; a parallelogram has 4 sides.

18. All; the opposite sides of a square are parallel.

19. Some; some parallelograms have four right angles.

20. No; triangles are not quadrilaterals.

21. $3\frac{1}{2}$ in.^{2}

22. B

23. D

24. Bottom: 36 in.^{2}
Back: 78 in.^{2}
Roof: 54 in.^{2}

Answers for Spiral Review, page 394

1. False; $3 + 7 \times (9 - 4) = 38$

2. False; $(4 + 3) \times 2 \times 5 = 70$

3. True

4. False; $4 + 6 \div (2 \times 3) + 4 = 9$

5. False; $(13 - 5) \div 4 + 2 = 4$

6. False; $38 \div (2 + 17) \times 4 = 8$

7. 10.9, 9.5, 9

8. $16\frac{7}{9}$, 15, 13

9. $15\frac{1}{3}$, 17, 12

10. $12\frac{7}{9}$, 10, 23

11. 15

12. 20

13. 6

14. 9 in.; 8 in.

15. 25 cm; 6 cm, 8 cm

16. $48

1. False

2. True

3. True

4. True

5. $\angle 3$, $\angle 5$, and $\angle 7$

6. $\angle 1$ and $\angle 4$, $\angle 1$ and $\angle 2$,
$\angle 2$ and $\angle 3$, $\angle 3$ and $\angle 4$,
$\angle 5$ and $\angle 6$, $\angle 5$ and $\angle 8$,
$\angle 6$ and $\angle 7$, $\angle 7$ and $\angle 8$

7. $125°$

8. $(-5, 2)$, $(-6, -2)$, $(0, 1)$

9. $(-1, 6)$, $(-3, 3)$, $(2, 2)$, $(4, 5)$

10. 3.00 m

11. 25 mm

12. 4.5 cm

13. 224 in.^{2}

14. 80 in.^{2}

Answers for Lesson 8.5, pages 397–399

Ongoing Assessment		
1. 35	**2.** 48	**3.** 90

Practice and Problem Solving

1. Answers vary.

2. B

3. C

4. A

5. The height is 40, not 42.
$\frac{1}{2}(22 + 30) \cdot 40 = 1040 \text{ cm}^2$

6. *Sample answer:* A parallelogram has 2 pairs of parallel sides, while a trapezoid has only 1 pair of parallel sides.

7. 9000 mm^2

8. 76.68 in.^2

9. 11.96 ft^2

10. 17.5 in.^2

11. 14.24 m^2

12. 114 cm^2

13. $18\frac{1}{16} \text{ mi}^2$

14.

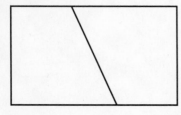

15.

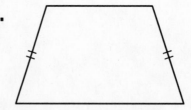

16.

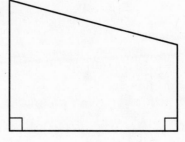

17. 2 mi

18. 20 cm

(continued)

19. 60 units^2

20. 500 units^2

21. C

22. D

23. Each side: 69 in.^{2}

    a. 69 in.^{2}

    b. The first method uses the formula for the area of a trapezoid. The second method uses two trapezoids to make a rectangle and then divides the area of the rectangle by 2.

Answers for Lesson 8.6, pages 401–404

Practice and Problem Solving

1. A

2. C

3. D

4. B

5. 92.86 ft^2

6. Answers vary.

7. 730.25 cm^2

8. 660.19 in.^{2}

9. 200.96 in.^{2}

10. 78.5 ft^2, 31.4 ft

11. 314 cm^2, 62.8 cm

12. 283.39 in.^{2}, 59.66 in.

13. 615.44 mi^2, 87.92 mi

14. 209.5 cm^2

15. 110.79 ft^2

16. 150.72 m^2

17. a

18. 7850 ft^2

19. 17,662.5 mi^2

20. 0.13 ft^2

21. 4.52 ft^2

22. 4.52 ft^2

23. 8.04 ft^2

24. B

25. C

26. 56.86 in.^{2}; Find the area of the rectangle and subtract the area of the circle.
362.86 in.^{2}; about 12.6 min

Answers for Spiral Review, page 404

1. $1, 3, 5, 7, 9$

2. $1\frac{2}{5}$ lb

3. D

4. B

5. A

6. C

7. Zero, 0

8. Positive, 14

9. Negative, -6

10. Negative, -8

11. Positive, 28

12. Zero, 0

13. $LM = 9$

14. $\angle A = 60°, AC = 40,$ $\angle F = 30°, FE = 28$

15. 25

16. 60%

17. 24

18. 800

Answers for Communicating About Mathematics, page 405

1. $34\frac{1}{6}$ feet

2. 405.93 feet^2. Subtract the area of the circle with diameter $25\frac{1}{2}$ feet from the area of the circle including the flattened grain.

3. 4069.44 feet^2

Answers for Lesson 8.7, pages 407–409

Ongoing Assessment		
1. 9.70	**2.** 5.29	**3.** 12.04

Practice and Problem Solving

1. $9^2 = 81, \sqrt{81} = 9$

2. No, because the square root of 20 cannot be written exactly as a decimal.

3. *Sample answer:* 121, 144, and 169

4. A

5. C

6. B

7. 5

8. 17

9. 1

10. 2.8

11. 9.1

12. 3.46

13. 7.75

14. 6.58

15. 9.7 yd

16. 7.07 yd

17. 7.75 yd

18. 1.73

19. 3.16

20. 8.66

21. 13.53

22. 9.06

23. 4.90

24. 16.97

25. 20.02

26. 1, 2, 3, 4, 5; the square root of the sum of the first n odd numbers equals n.

27. $a = 256, b = 16, c = 4, d = 2$

28. 8 units

29. 32 units^2, $\approx$ 5.7 units

30. 312 ft^2

31. C

32. B

(continued)

33. a. 10 in.; No, the bottom of the birdhouse has sides of 6 in.

   b. Back: 13 in. by 10 in.
Bottom: 10 in. by 10 in.
Roof: 10 in. by 12.5 in.
Front: 10 in. by 10 in.
Sides: The bases of the trapezoid remain at 10 in. and 13 in. The height of the trapezoid changes from 6 in. to 10 in.

Answers for Lesson 8.8, pages 413–415

Ongoing Assessment		
1. 7.07	**2.** 25	**3.** 50

Practice and Problem Solving

1. Right, legs, hypotenuse

2. 225, 400; 625; 25

3. 900, 2500; 1600; 40

4. DE and EF

5. DF

6. 5 in.

7. Obtuse

8. Right, m

9. Acute

10. Right, 20 ft

11. Right, 5.66 cm

12. Obtuse

13. 39.05

14. 73.82

15. 114.13

16. 8 m

17. 36 in.

18. 14.14 yd

19. 10.11 ft

20. A

21. C

22. a. 2.29 in.

    b. The slanted portion of the side of the birdhouse is $\sqrt{6^2 + 3^2} \approx 6.71$. Subtract this from the 9 in. portion of the roof.

Answers for Chapter Review, pages 417–419

1. $\angle 3 = 35°$, $\angle 7 = 90°$,
   $\angle 8 = 35°$, $\angle 9 = 55°$

2. $(0, -6)$, $(3, -3)$, $(5, -2)$,
   $(4, -8)$

3. $(3, 0)$, $(6, 3)$, $(8, 4)$, $(7, -2)$

4. $(-2, 4)$, $(1, 7)$, $(3, 8)$, $(2, 2)$

5. 34.54 ft

6. 87.92 in.

7. 8 m

8. 15 units^2

9. 8 units^2

10. 25 units^2

11. 7.5 units^2

12. 15 units^2

13. 54 units^2

14. 153.86 ft^2

15. 113.04 in.^{2}

16. 11

17. 5.8

18. 4.12

19. 9.49

20. Obtuse

21. Right, 26

22. Right, 6

Answers for Chapter Assessment, page 420

1. Vertical
2. Supplementary
3. $82°$
4. $98°$
5. 28.26 in.^{2}
6. 73.7 m^2
7. 86.25 ft^2
8. 18.84 in.
9. 3 units to the right, 18 units up
10. $(-1, 24), (2, 6), (20, 4)$
11. 4.12

12. 6.48
13. 12.96
14. 14.14
15. 2 in.
16. 16 in.^{2}
17. 20 in.^{2}
18. 5 in.
19. 24 in.^{2}
20. The side length of a square is the square root of the area of the square. Answers vary.

Answers for Standardized Test Practice, page 421

1. B
2. B
3. D
4. D
5. B
6. B

7. B
8. C
9. B
10. A
11. C
12. D

Answers for Lesson 9.1, pages 429–431

Ongoing Assessment

1. 7 faces, 10 vertices, 15 edges

2. 8 faces, 12 vertices, 18 edges

The sum of the number of faces and the number of vertices is 2 more than the number of edges.

Practice and Problem Solving

1. No, faces are not polygons.

2. Yes, pentagonal pyramid

3. Yes, triangular prism

4. No, sides are not polygons.

5. a: face, b: vertex, c: edge

6. Number of faces + Number of vertices = Number of edges + 2

7. Answers vary.

8. Answers vary.

9.

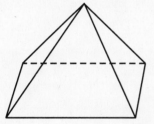

10.

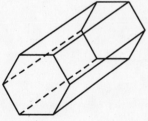

11.

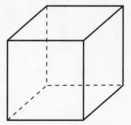

12. No

13. Yes, 5

14. No

15. Yes, 6

(continued)

16. Yes, a triangular pyramid

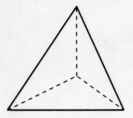

17. Net C. Net A has only 5 sides instead of 6. Net B has 2 sides that overlap. Net C forms a box with 6 sides.

18. Faces: 6, edges: 12, vertices: 8

19. B

20. C

21. B; faces: 6, vertices: 8, edges: 12; yes

22. C; faces: 9, vertices: 9, edges: 16; yes

23. A; faces: 8, vertices: 6, edges: 12; yes

Answers for Lesson 9.2, pages 433–435

Ongoing Assessment

1. 3 cans of green paint

2. 4 cans of purple paint

Practice and Problem Solving

1. Top: 160 in.^{2},
Sides: 100 in.^{2},
Ends: 12 in.^{2}

2. 384 in.^{2}

3. A

4. *Sample answer:*

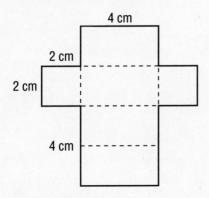

5. A; 52 cm^2

6. 24 ft^2

7. $13\frac{1}{8}$ in.^{2}

8. 152 cm^2

9. 12 inches

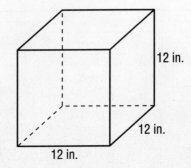

10. 1056 in.^{2}

11. 204.8 cm^2

12. 1240 in.^{2}

13. A. Each of the 6 faces has an area that is about $\frac{1}{4}$ in.^{2}

14. B. Each of the 2 largest faces has an area that is about 21 ft^2, while the remaining 4 faces together have an area that is less than 2 ft^2.

15. 402 in.^{2}

16. 214 in.^{2}

17. D

18. B

19. 268 in.^{2}

Answers for Spiral Review, page 436

1. $\frac{1}{12}$

2. $4\frac{4}{5}$

3. 12

4. $\frac{7}{20}$

5. $4\frac{1}{8}$

6. $12\frac{1}{4}$

7. $2\frac{1}{9}$

8. $\frac{104}{135}$

9. *Sample answer:*

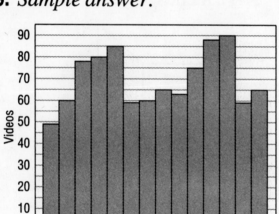

A pictograph or a bar or line graph

10. Byron

11. 24

12. 20

13. 125%

14. 245

15. 95

16. 75%

Answers for Lesson 9.3, pages 439–441

Ongoing Assessment

1.

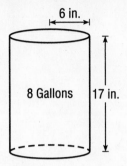

6 in.

8 Gallons   17 in.

2. 6028.8 cm^2

3. The larger can uses 4 times as much metal as the smaller can.

Practice and Problem Solving

1. *a*: base, *b*: circumference,
   c: height, *d*: radius,
   e: lateral surface

2. Yes

3. D

4. C

5. A

6. B

7. Answers vary.

8. Yes

9. No

10. No

11. Yes

12. 244.92 m^2

13. 1155.52 ft^2

14. 502.40 m^2

15. A; 4873.3 mm^2

16. D; 14.8 cm^2

17. B; 60.83 in.^{2}

18. C; 626.4 mm^2

19. *Sample answer:*

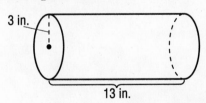

3 in.

13 in.

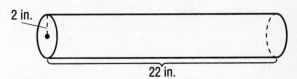

2 in.

22 in.

(continued)

Answers for Lesson 9.3, pages 439–441 (cont.)

20. b
   a. Check: 504.8 cm^2,
   b. Check: 401.9 cm^2

21. No; sack: 2060 in.^{2}, bag: 5172 in.^{2}

22. B

23. A

24. 7.85 in.^{2}, 7.85 ft^2

Answers for Lesson 9.4, pages 443–446

Ongoing Assessment
1. No; by changing the stack, at least one view will change.

Practice and Problem Solving

1.
Top

Front   Side

2.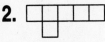
Top

Front   Side

3.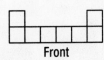
Top

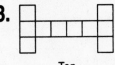
Front   Side

4. Answers vary.

5.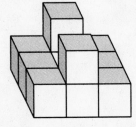

6. C
7. A
8. B
9. Triangular prism
10. Cube
11. Rectangular prism
12. C
13. A
14. B
15. B
16. A
17. *Sample answer:*

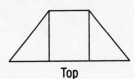
Top

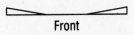 
Front   Side

Answers vary.

Answers for Spiral Review, page 446

1. 6

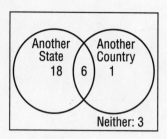

2. 7

3. 3.25

4. 1

5.

5   10 14   27   49

6. 58% decrease

7. 187% increase

8. ∠1 and ∠3, ∠2 and ∠4, ∠5 and ∠7,
∠6 and ∠8, ∠9 and ∠11, ∠10 and ∠12

Answers for Mid-Chapter Assessment, page 447

1.

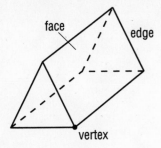

face, edge, vertex

2. Cylinder

3. Triangular prism

4. Pentagonal pyramid

5. *Sample answer:*

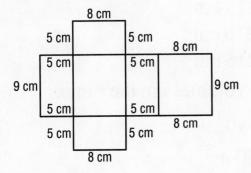

6. *Sample answer:*

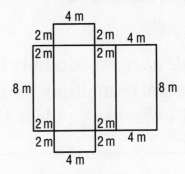

7. *Sample answer:*

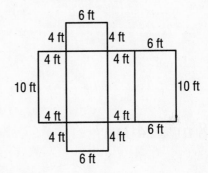

8. 672 in.^{2}

9. 459 ft^2

10. 42.45 in.^{2}

11. Triangular prism

12. 310.5 in.^{2}

Answers for Lesson 9.5, pages 451–453

Practice and Problem Solving

1. B

2. C

3. A

4. 248 m^2, 240 m^3

5. x^3

6. 0.384 m^3

7. 2496 in.^3

8. 72 ft^3

9. $334{,}540{,}800 \text{ ft}^3$

10. 2 in.

11. 5 m

12. 10 mm

13. **a.** 216 cm^3
    b. 216 cm^3
    c. 216 cm^3
    The volumes are the same.

14. 384 yd^3

15. 61.50 m^3

16. 7470.4 ft^3

17. 800 in.^3

18. C

19. B

20. 25 in.^3, 25 ft^3

Answers for Lesson 9.6, pages 455–458

Ongoing Assessment
Cylinder #2; squaring the radius 6 results in a greater volume than squaring the radius 4.

1. Cylinder #1: 301.44 cm^3

2. Cylinder #2: 452.16 cm^3

Practice and Problem Solving

1. A; because A gives you the amount of juice, while B gives you the lateral surface area.

2. The shorter cylinder; shorter: 82 in.^{3}, taller: 63 in.^{3}

3. 4.5 ft

4. 63.6 ft^2

5. 190.8 ft^3

6. 1582.6 mm^3

7. 1081.6 cm^3

8. 1139.8 yd^3

9. $h \approx 6$ m

10. $r \approx 9$ m

11. 8 entire walls

12. *Sample answer:*

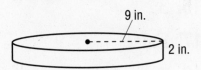

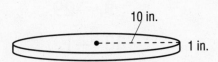

Volume: 509 in.^{3}
Surface area: 622 in.^{2}

Volume: 314 in.^{3}
Surface Area: 691 in.^{2}

(continued)

13.

Height	Radius	Volume
1 unit	1 unit	3.14 units^3
1 unit	2 units	12.56 units^3
1 unit	3 units	28.26 units^3
1 unit	4 units	50.24 units^3

When you double the radius, you multiply the volume by 4; when you triple the radius, you multiply the volume by 9; when you quadruple the radius, you multiply the volume by 16, and so on.

14. 6556 in.^{3} Subtract the volume of the tree without the bark from the volume of the tree.

15. a. 7.95 cm^3

    b. 91,354 pencils

16. C

17. D

18. D

19. 1.96 ft^3

Answers for Spiral Review, page 458

1. 32

2. 50

3. 64

4. 81

5. 5

6. 2

7. 19

8. -12

9. -29

10. -13

11. 16

12. 4

13. 12

14. 2

15. 7

16. 3

17. False

18. True

19. $3.44

20. 52.38 ft^2

21. 37.68 m^2

22. 557 yd^2

Answers for Communicating About Mathematics, page 459

1. State teams: 312 ft^2
International teams: 600 ft^2
High school teams: 128 ft^2

2. 24 lb; $\frac{1}{72}$ lb; the weight of 1 in.^{3} of snow is $\frac{1}{1728}$ of
1 ft^3 of snow.

3. 28,800 lb

4.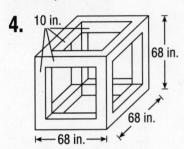

5. $248,832 \text{ in.}^3$; about 62.9 in.

Answers for Lesson 9.7, pages 463–465

Ongoing Assessment

1. Length: $\frac{32}{15}$ ft ≈ 2.13 ft

Width: $\frac{4}{9}$ ft ≈ 0.44 ft

Height: $\frac{22}{45}$ ft ≈ 0.49 ft

Practice and Problem Solving

1. C

2. A

3. B

4. Yes. Since the length, width, and height of each cube are the same, the ratios of their corresponding dimensions must be equal; also, any two cubes have the same shape.

5. Their corresponding dimensions are not proportional.

$$\frac{4 \text{ ft}}{15 \text{ in.}} \neq \frac{4 \text{ ft}}{6 \text{ in.}}$$

6. *Sample answer:*

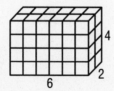
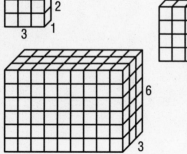

7. Yes

8. No

9. Yes

10. Yes

11. 484 units^2, 1936 units^2; false

12. 62.5 ft, 18 in.

13. 45 cm, 50 m

14. 30 m, 6.25 m

15. 8 cm, 1.25 cm

16. a, c, d;
volume of closet: 120 ft^3;
volumes of a, c, and d:
240 ft^3

17. D

18. B

19. Answers vary.

Answers for Chapter Review, pages 467–469

1. Faces: 7; pentagonal prism; edges: 15; vertices: 10

2. Faces: 7; hexagonal pyramid; edges: 12; vertices: 7

3. Faces: 5; triangular prism; edges: 9; vertices: 6

4. Faces: 4; triangular pyramid; edges: 6; vertices: 4

5. *Sample answer:*

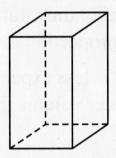 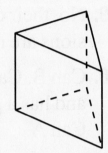

A rectangular prism has parallel bases that are rectangles, while a triangular prism has parallel bases that are triangles.

6.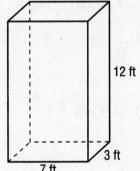

282 ft^2

7. 432 in.^{2}

8. 368 ft^2

9. 226.1 cm^2

10. 340.7 in.^{2}

11. 9

12.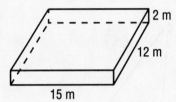

360 m^3

13. 2,112,000 ft^3

14. 528,000 ft^3

15.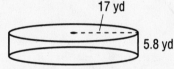

5263.27 yd^3

16. b. Squaring 3 has a larger effect than squaring $2\frac{11}{16}$.

17. $x = 4.95$ ft, $y = 5$ ft

18. The other boxes should be cubes with side lengths larger than the next smallest box.

19. Similar

20. Not similar

Answers for Chapter Assessment, page 470

1.–3. *Sample answers*

1.

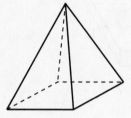

2.

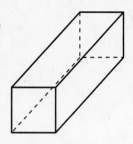

3.

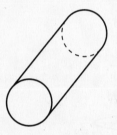

4. 199.1 cm^2, 150.8 cm^3

5. 558 ft^2, 770 ft^3

6. 1879.3 in.^{2}, 6231.3 in.^{3}

7. 231.6 m^2, 217.7 m^3

8. Yes, their corresponding dimensions are proportional.

9. No, their corresonding dimensions are not proportional.

10. Can B. Can B is less expensive and has a greater volume than Can A.

11. Can A: 28.6 in.^{2}
Can B: 26.7 in.^{2}
Lateral surface area

12. 9043 cm^3

13. 1140 in.^{2}, 14.6 gallons

Answers for Standardized Test Practice, page 471

1. B

2. C

3. A

4. C

5. D

6. B

7. B

8. B

Answers for Cumulative Review, pages 472 and 473

1. $0.35, \frac{7}{20}$

2. $0.0284, \frac{71}{2500}$

3. $0.004, \frac{1}{250}$

4. $2.5, \frac{5}{2}$

5. 25%

6. 8%

7. 62.5%

8. 72.5%

9. 63

10. 22%

11. 8%

12.

13. $110.25, $460.25

14. $67.50, $1267.50

15. $6.75

16. 22.5%

17. $35

18. $1.25

19. 18.84 ft

20. 78.5 cm^2

21. 10 m

22. 26 units, 27 units^2

23. 50.6 units, 103.8 units^2

24. 20 units, 25 units^2

25. 10 in.

26. 17.0 cm

27. 12 ft

28. Triangular prism, 120 in.^{2}

29. Rectangular prism, 232 ft^2

30. Cylinder, 175.8 cm^2

31. Cylinder

32. Rectangular prism

33. The cylindrical silo can be built with less material and holds more feed.

34. $a = 10$ cm, $b = 6$ cm

35. $a = 20$ in., $b = 60$ in.

Answers for Lesson 10.1, pages 481–483

> **Ongoing Assessment**
>
> **1.** False. The absolute value of zero is not positive.
>
> **2.** True. Opposite numbers are the same distance from zero.
>
> **3.** True. All other numbers lie some distance from zero and, so, have a positive absolute value.

Practice and Problem Solving

1. G

2. B

3. D and F

4. False, the absolute value of point C is 5.

5. True

6. True

7. False; point G's absolute value is 4, point A's absolute value is 8, and $8 > 4$.

8.
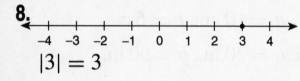
$|3| = 3$

9.

$|-2| = 2$

10.
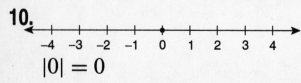
$|0| = 0$

11.
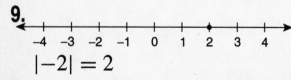
$|-15| = 15$

12. $|6|$

13. $|4.5|$

14. $|-4|$

15. $|-6.5|$

16. 5

17. 6

18. 8

19. 0

20. 22.6

21. 2.7

22. 3.7

23. 0.25

24. 7.3

(continued)

Answers for Lesson 10.1, pages 481–483 (cont.)

25. 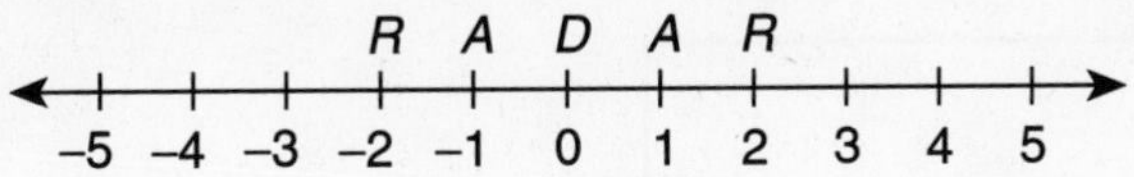
R A D A R
–5 –4 –3 –2 –1 0 1 2 3 4 5

26. $=$

27. $<$

28. $>$

29. $=$

30. $>$

31. $<$

32. \$2.05 under budget

33. \$1.25 over budget

34. \$1.75 over budget

35. Clothes; $|2.05| > |-1.25|$ and $|2.05| > |-1.75|$

36. C

37. C

38. 1036 ft

Answers for Lesson 10.2, pages 485–487

Practice and Problem Solving

1.–3. a is a negative number and b is a positive number.

1. Zero; a and b are the same distance from zero, so they are opposites.

2. Positive; b is farther from zero than a is, so b has the greater absolute value.

3. Negative; a is farther from zero than b is, so a has the greater absolute value.

4.–7. *Sample answers*

4. 1 and -9

5. -1 and 18

6. 1 and -18

7. 1 and -1

8. $-6 + (-8) = -14$

9. -14; add 50 and -14 to get 36.

10. Negative, -16

11. Zero, 0

12. Positive, 2

13. Negative, -5

14. 1

15. -8

16. -3

17. -14

18. -18

19. -39

20. -9

21. 7

22. 9

23. 7

24. 26

(continued)

Answers for Lesson 10.2, pages 485–487 (cont.)

25. -9

26. -12

27. 50

28. 2

29. 5

30. 2

31.–34. *Sample answers*

31. -1 and 2

32. -1 and 10

33. 1 and -3

34. 1 and -16

35. At the end of the month. $40 + 10 + 20 + 25 = 95$, $(-25) + (-20) + (-35) = -80$, $95 + (-80) = 15$, $\$15$ more.

36. $a = 7, b = -7, c = 0, d = 4$

37. Joel won.
Juan: $3 + 0 + 2 + (-2) + 1 + (-1) + 4 + (-2) + 0 = 5$
Joel: $-2 + 1 + (-1) + (-1) + 2 + 3 + 0 + (-1) + (-1) = 0$
Sara: $4 + 2 + (-2) + (-1) + 0 + 0 + (-1) + 2 + (-2) = 2$
$0 < 2$ and $0 < 5$

38. D

39. C

40. -3730 meters

Answers for Lesson 10.3, pages 489–491

Ongoing Assessment

1. 12, Cairo is 12 hours later than Honolulu.

2. 8, Ponta Delgada is 8 hours later than Anchorage.

3. 7, Greenwich is 7 hours later than Denver.

Practice and Problem Solving

1. C, -2

2. A, -8

3. B, 8

4. B, $19°C$

5. $45

6. $15

7. $5 + (-18); -13$

8. $0 + (-12); -12$

9. $-3 + (-19); -22$

10. $-8 + (-27); -35$

11. $-12 + (-25); -37$

12. $0 + 36; 36$

13. $-5 + 27; 22$

14. $-11 + 19; 8$

15. 11

16. -18

17. -4

18. -15

19. -22

20. 45

21. 1

22. -56

23. -8

24. -22

25. 18

26. -30

27. 1, -6

28. 6, 13

29. $-14, -7$

30. 9, 2

31. $-6, 1$

32. $-10, -3$

33. 8, 1

34. $-4, -11$

(continued)

35. C; −31

36. A; 31

37. B; −11

38. Positive. $0 - x = 0 + (-x) = -x$; if the opposite of x is negative, then x is positive.

39. Positive. $0 - x = 0 + (-x) = -x$; if x is negative, then the opposite of x is positive.

40. 2 min

41. 9 min

42. 6 min

43. Brand B; it lasts longer.

44. B

45. Mount St. Helens

Answers for Lesson 10.4, pages 493–496

Ongoing Assessment

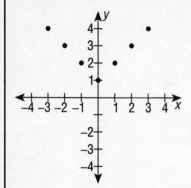

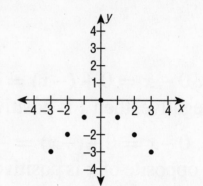

The points lie on the sides of a right angle whose vertex is at (1, 0).

The points lie on the sides of a right angle whose vertex is at (0, 0).

Practice and Problem Solving

1. Answers vary.

2. 185

3. No, $220 - 65 \neq 145$.

4. B

5. C

6. A

7. $-17, -14, -11, -8, -5$

8. $4, 1, -3, -6, -10$

9. $-5, -8, -10, -13, -17$

10. $(-2, 8), (-1, 7), (0, 6), (1, 5), (2, 4)$

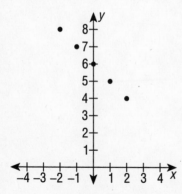

The points lie on a line that slopes down to the right.

(continued)

11. $(-2, -7), (-1, -6), (0, -5), (1, -4), (2, -3)$

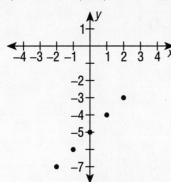

The points lie on a line that slopes up to the right.

12.

x	−4	−3	−2	−1	0	1	2	3	4
y	5	6	7	8	9	10	11	12	13

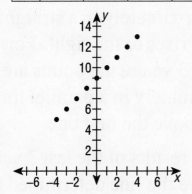

The points lie on a line that slopes up to the right.

13.

x	−4	−3	−2	−1	0	1	2	3	4
y	−15	−14	−13	−12	−11	−10	−9	−8	−7

The points lie on a line that slopes up to the right. *(continued)*

14.

x	−4	−3	−2	−1	0	1	2	3	4
y	−1	−2	−3	−4	−5	−6	−7	−8	−9

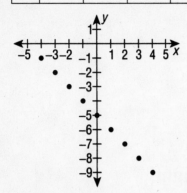

The points lie on a line that slopes down to the right.

15.

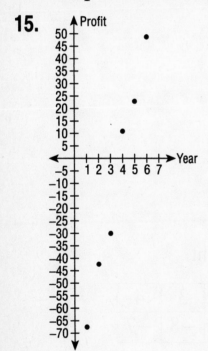

16. For the first 3 years, the points are approximately in a straight line that rises to the right. For the next 3 years, the points are approximately in a parallel line that is above the first one.

17. Yes; the results of the last 3 years are above the line determined by the results of the first 3 years.

18.

x (cm)	1	2	3	4	5	6	7	8
P (cm)	14	16	18	20	22	24	26	28

(continued)

19.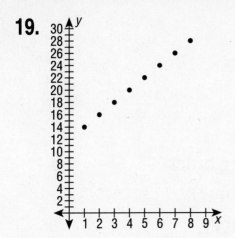

The points lie on a line that slopes up to the right; or as the width increases by 1 cm, the perimeter increases by 2 cm.

20. a.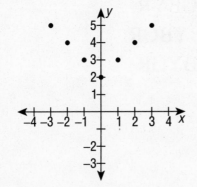

b.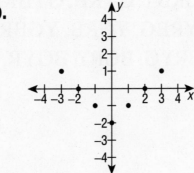

Alike: The points lie on the sides of a right angle.
Different: Their vertices are different; the vertex of **a** is $(0, 2)$ and the vertex of **b** is $(0, -2)$.

21. D   **22.** B

23.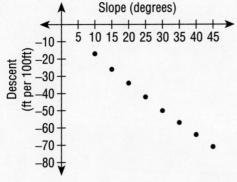

The greater the slope, the greater the descent per 100 ft.

Answers for Spiral Review, page 496

1.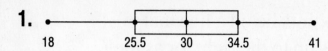

2.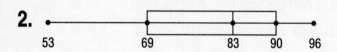

3. The turtles may finish in any of the following orders (R = red, G = green, Y = yellow, and B = blue):

RGYB, RGBY, RYGB, RYBG, RBYG, RBGY,
GRYB, GRBY, GYRB, GYBR, GBRY, GBYR,
YRGB, YRBG, YGRB, YGBR, YBRG, YBGR,
BRGY, BRYG, BGRY, BGYR, BYRG, BYGR

4. $\frac{1}{12}$

5. 150

6. $40.28

7. $262.50

8. 24 cm

9. 5 in.

10. 8.49 m

Answers for Mid-Chapter Assessment, page 497

1. False; $|3| = 3$, 3 is not the opposite of 3.

2. True

3. False. To subtract an integer you add its opposite, and the opposite of a negative integer is always a positive integer; the sum of two positive integers is always a positive integer.

4. 1.5

5. 4

6. 20

7. 0

8. 8

9. 5

10. 19

11. -23

12. -11

13. -12

14.

x	-3	-2	-1	0	1	2	3
y	7	6	5	4	3	2	1

$(-3, 7), (-2, 6), (-1, 5), (0, 4), (1, 3), (2, 2), (3, 1)$

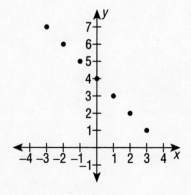

(continued)

15.

x	-3	-2	-1	0	1	2	3
y	0	-1	-2	-3	-2	-1	0

$(-3, 0)$, $(-2, -1)$, $(-1, -2)$, $(0, -3)$, $(1, -2)$, $(2, -1)$, $(3, 0)$

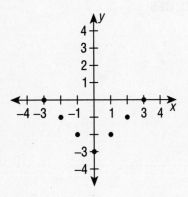

16. 800,000 B.C. is more recent. 800,000 B.C. means 800,000 years before zero or $-800,000$, while 1,600,000 B.C. means 1,600,000 years before zero or $-1,600,000$; 800,000 B.C. is closer to today's date.

17. 1260 years

18. 10 B.C.: -2010; 1250 A.D.: -750

Answers for Lesson 10.5, pages 501–503

Ongoing Assessment

1. $-2 \cdot (-3) \cdot (-1) = 6 \cdot (-1)$ Multiply –2 by –3.

$\qquad\qquad\qquad\quad\; = -6$ Multiply 6 by –1.

2. $-3 \cdot (-3) \cdot (-3) \cdot (-3) = 9 \cdot (-3) \cdot (-3)$ Multiply –3 by –3.

$\qquad\qquad\qquad\qquad\quad\; = -27 \cdot (-3)$ Multiply 9 by –3.

$\qquad\qquad\qquad\qquad\quad\; = 81$ Multiply –27 by –3.

3. $3 \cdot (-4) \cdot (-10) \cdot 2 = -12 \cdot (-10) \cdot 2$ Multiply 3 by –4.

$\qquad\qquad\qquad\qquad\quad = 120 \cdot 2$ Multiply –12 by –10.

$\qquad\qquad\qquad\qquad\quad = 240$ Multiply 120 by 2.

Practice and Problem Solving

1. Positive. The product of two negative numbers is positive.

2. Zero. The product of any number and zero is zero.

3. Negative. The product of a positive number and a negative number is negative.

4. Positive. The product of two positive numbers is positive.

5. 28

6. −36

7. 40

8. −3, 18

9. Zero. Any number subtracted from itself is zero.

(continued)

10. Negative. The sum of two negative numbers is negative.

11. Positive. The product of two positive numbers is positive.

12. Zero. The product of any number and zero is zero.

13. Negative. The product of a positive number and a negative number is negative.

14. Positive. The product of two negative numbers is positive.

15. 56

16. 0

17. -9

18. 26

19. -16

20. 72

21. 8

22. -9

23. -1

24. 0

25. -13

26. 3

27. negative sign; -224

28. negative sign; -172

29. negative sign; $-11,340$

30. 42

31. -4

32. -216

33. 0

34. -6

35. 0

36. $(-1)(-7)(9); 63$

37. $(5)(-6)(0); 0$

38.–41. *Sample answers*

38. $0 \cdot 5 = 0; -14 \cdot 0 = 0$

39. $3 \cdot (-4) = -12;$
$-3(4) = -12$

40. $8(-5) = -40;$
$-20 \cdot 2 = -40$

41. $(-5)(-7) = 35;$
$5 \cdot 7 = 35$

(continued)

Answers for Lesson 10.5, pages 501–503 (cont.)

42. $39, −$25, −$3.50, $19, −$13.50; $16

43. C

44. D

45.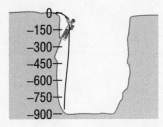

−150, −300, −450, −600, −750, −900

Answers for Lesson 10.6, pages 505–508

Ongoing Assessment

1. -238 ft, subtract the sum of the five readings from $6(-254)$ to get the sixth reading.

Practice and Problem Solving

1. False; for example:
   $3 \div (-3) = -1$, not 1.

2. False; for example:
   $3 \div 3 = 1$, not -1.

3. Cannot divide by zero.

4. Zero divided by any nonzero number is zero.

5.

−16	−11	5	−32	−20

50	29	18 $+(-1) = $ −18 $+(-3) = $ 6

↑ ÷2   ÷2 ↓

108 $+(-3) = $ −36 $+(-1) = $ 36   22   3

6.

−85	75	30 $+(-3) = $ −10 $+(-1) = $ 10

↑ ÷ (−2)

−240 $+(-1) = $ 240 $+(-4) = $ −60   −55   15

↑ ÷ (−2)

480	100	135	205	40

7. Negative

8. Negative

9. Positive

10. Positive

11. -7

12. -3

13. 3

14. 0

15. -13

16. -25

17. 0

18. 44

19. Not possible

20. 19

21. 5

22. Not possible

23. $-3°$

24. $-\dfrac{5}{6}$

25. -9 ft

26. $-\$1$

27. $\$1.83$ million

28. -1.2 miles, $-0.\overline{3}$ miles, -0.5 miles, -1.25 miles

(continued)

Answers for Lesson 10.6, pages 505–508 (cont.)

29. $2, -2$

30. $1, -1$

31. $-3, 3$

32. $-6, 6$

33. -35

34. -9

35. -3

36. 15

37. -2

38. $-1\frac{1}{2}$ lb

39. A

40. C

41. -6.5 miles

Answers for Spiral Review, page 508

1. $2m + 18.50$

2. $\dfrac{\$5}{1 \text{ day}}$

3. $\dfrac{55 \text{ miles}}{1 \text{ hour}}$

4. 2 units to the left and 4 units down

5. 3 units to the right and 1 unit up

6. $169 \text{ cm}^2, 132 \text{ cm}^3$

7. $351.68 \text{ ft}^2, 502.4 \text{ ft}^3$

Answers for Communicating About Mathematics, page 509

1. 35 feet; $|-35| = 35$

2.

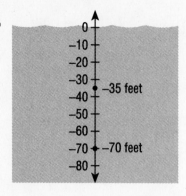

$|-35 - (-70)| = 35$ feet

3. 10,344 feet. No. Subtracting a smaller number from a larger number gives a positive answer.

4. $-12{,}000$ feet; multiply the depth (-6000 feet) by 2.

5. $-13{,}124$ feet

6. $-13{,}447$ feet

Answers for Lesson 10.7, pages 511–513

Ongoing Assessment

1.

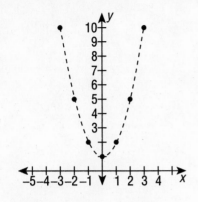

The points form a
U-shaped pattern.

2.

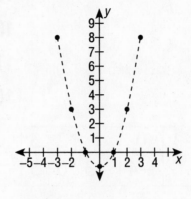

The points form a
U-shaped pattern.

Practice and Problem Solving

1. $(-3)^4$; 81

2. $(-2)^5$; -32

3. A negative integer raised to an
odd power is negative, while
a negative integer raised to an
even power is positive.

4. D

5. B

6. C

7. A

8. $-27, -8, -1, 0, 1, 8, 27$

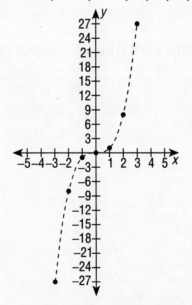

The points lie on a curved line
that rises to the right; when the
curved line passes through the
origin, it is nearly horizontal.

(continued)

Answers for Lesson 10.7, pages 511–513 (cont.)

9. Answers vary.

10. 81

11. −64

12. 8

13. 64

14. 70

15. −59

16. 24

17. 10

18. 33

19.

x	−3	−2	−1	0	1	2	3
y	5	0	−3	−4	−3	0	5

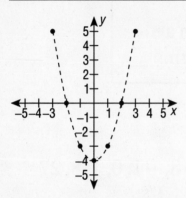

The points form a U-shaped pattern.

20.

x	−3	−2	−1	0	1	2	3
y	18	8	2	0	2	8	18

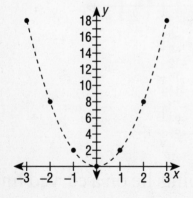

The points form a U-shaped pattern.

(continued)

21.

x	−3	−2	−1	0	1	2	3
y	−9	−4	−1	0	−1	−4	−9

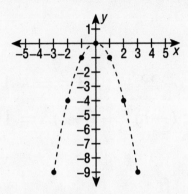

The points form an upside-down U-shaped pattern.

22.

x	−3	−2	−1	0	1	2	3
y	−6	−1	2	3	2	−1	−6

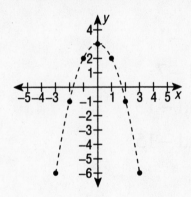

The points form an upside-down U-shaped pattern.

23. $=$

24. $<$

25. $>$

26. $=$

27. $<$

28. $>$

29. $=$

30. $<$

31. $=$

(continued)

Answers for Lesson 10.7, pages 511–513 (cont.)

32.

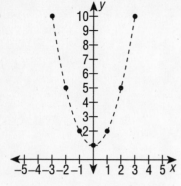

The points form a U-shaped pattern.

33.

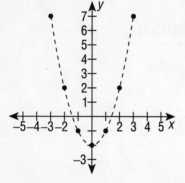

The points form a U-shaped pattern.

34. 6250

35. −5375

36. −3915

37. True

38. False; $-2^3 - (-3)^2 = -17$

39. False; $(-1)^6 + (-2)^4 = 17$

40. True

41. a. 960 ft

b. 624 ft

c. 0 ft

42. D

43. D

44. 1 and −1

45. 7 and −7

46. 13 and −13

47. 15 and −15

Answers for Spiral Review, page 514

1. 20

2. 21

3. 3

4. 7

5. 11.66

6. 3.3

7. 5.2

8. 30%

9. 79.2°

10. 700

Answers for Lesson 10.8, pages 517–519

Ongoing Assessment
1. 8×10^9; $8 \times 10^9 = 80 \times 10^8$, $80 > 9$
2. 5×10^6; $5 \times 10^{-7} = 0.5 \times 10^{-6}$, $5 > 0.5$
3. 4×10^{-3}; $3 \times 10^{-4} = 0.3 \times 10^{-3}$, $4 > 0.3$

Practice and Problem Solving

1. C

2. A

3. B

4. **a.** No, there is no power of 10.
   b. No, 0.35 is not between 1 and 10.
   c. Yes

5. 5.28×10^3

6. 3.6×10^3

7. 1×10^{-3}

8. 100,000

9. 10,000,000,000

10. $\frac{1}{1,000,000}$

11. $\frac{1}{1,000,000,000,000}$

12. 5.4×10^5

13. 1.205×10^7

14. 6.2×10^{-2}

15. 3.5×10^{-4}

16. 1.45×10^{11}

17. 6.67×10^{-7}

18. 1.4×10^{-3}

19. 5×10^{-4}

20. 1.6×10^9

21. 1.6×10^7

22. 13.45 is not between 1 and 10.
    1.345×10^5

23. The exponent should be -5, not 5.
    5.6×10^{-5}

24. $<$

25. $<$

26. $>$

27. $<$

28. $>$

29. $>$

(continued)

Answers for Lesson 10.8, pages 517–519 (cont.)

30. B; A is too small a number.

31. A; A is a very small number.

32. A

33. D

34. A. VEI 3
    B. VEI 6
    C. VEI 0
    D. VEI 4

Answers for Chapter Review, pages 521–523

1. 75

2. 4.7

3. 0

4. March, April, and June. During January and February the company lost money, then during the next four months it made a profit. For the 6 months, the income was $335 more than the expenses.

5. 13

6. −14

7. 0

8. −14

9. −4

10. −28

11. 19

12. −5

13. −5, −4, −3, −2, −1, 0, 1

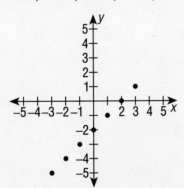

The points lie in a straight line.

(continued)

Answers for Chapter Review, pages 521–523 (cont.)

14.

x	−3	−2	−1	0	1	2	3
y	5	4	3	2	1	0	−1

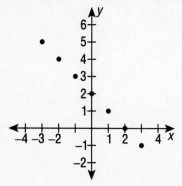

15.

x	−3	−2	−1	0	1	2	3
y	−4	−3	−2	−1	0	1	2

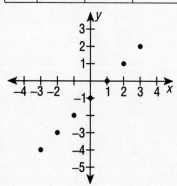

16.

x	−3	−2	−1	0	1	2	3
y	0	−1	−2	−3	−2	−1	0

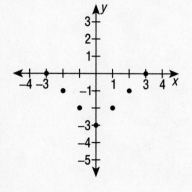

(continued)

Passport to Mathematics Book 2

17.

x	−3	−2	−1	0	1	2	3
y	7	6	5	4	5	6	7

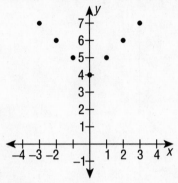

18. 72

19. −35

20. −3

21. 54

22. 42

23. −44

24. −12°F

25. 6

26. −5

27. −5

28. 15

29. −804.6

30. −18

31. −20

32. −225

33. 3.4×10^4

34. 6.1×10^{-3}

35. 7.2×10^{-6}

36. 5.7×10^8

37. 1×10^3

38. 8.0×10^{10}

1. False. The absolute value of a negative number is its distance from zero, which is positive.

2. True

3. -5

4. -10

5. -9

6. 9

7. 0

8. -49

9. 36

10. 9

11. -7

12. 47

13. 256

14. -81

15. $=$

16. $=$

17. $<$

18. $-1, 0, 1, 2, 3, 4, 5$

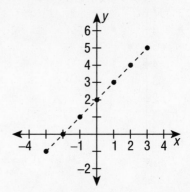

The points lie on a line that slopes up to the right.

19. $4, 3, 2, 1, 0, 1, 2$

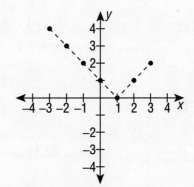

The points lie on the sides of a right angle whose vertex is at $(1, 0)$.

(continued)

Answers for Chapter Assessment, page 524 (cont.)

20. 2.2×10^{-2}

21. 6.0×10^{8}

22.

Week (x)	1	2	3	4	5	6
Pay (y) in $	1	2	4	8	16	32

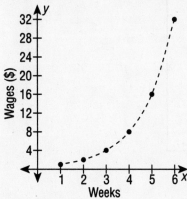

23. 8th week

24. Yes; during the 21st week alone, you will have earned $1,048,576.

Answers for Standardized Test Practice, page 525

1. C

2. B

3. B

4. C

5. B

6. D

7. A

8. C

9. B

10. A

Passport to Mathematics Book 2　　　　*Answer Masters* **205**

Copyright © McDougal Littell Inc. All rights reserved.

Answers for Lesson 11.1, pages 531–533

Ongoing Assessment

1.

	R	R
R	RR	RR
W	RW	RW

Red and pink parents:

Probability of red $= \frac{1}{2}$

Probability of pink $= \frac{1}{2}$

2.

	R	R
W	RW	RW
W	RW	RW

Red and white parents:

Probability of pink $= 1$

Practice and Problem Solving

1. Answers vary.

2. $\frac{2}{5}$

3.

11	21	31	41	51	61
12	22	32	42	52	62
13	23	33	43	53	63
14	24	34	44	54	64
15	25	35	45	55	65
16	26	36	46	56	66

$\frac{1}{4}$

4. Answers vary.

5. $\frac{61}{100}$

6. Both parents are round;
offspring: probability of round $= \frac{3}{4}$
probability of wrinkled $= \frac{1}{4}$

7. One parent is round, one is wrinkled;
offspring: probability of round $= \frac{1}{2}$
probability of wrinkled $= \frac{1}{2}$

(continued)

8. One parent is round, one is wrinkled; offspring: probability of round $= 1$

9. Both parents are round; offspring: probability of round $= 1$

10. 2 sectors are red, 1 is yellow, 5 are green, and 4 are blue.

11. $\frac{1}{12}$

12. $\frac{39}{200}$

13. $\frac{1}{2}$

14. Russian

15. C

16. B

17. $\frac{1}{5}$

Answers for Lesson 11.2, pages 537–539

Ongoing Assessment

1. 210

2. There are now 10 waist sizes, 7 lengths, and 3 styles, so $10 \times 7 \times 3 = 210$ different types of jeans.

Practice and Problem Solving

1. Answers vary.

2. 676, $26 \times 26 = 676$

3.

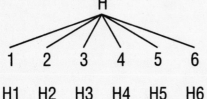

H   T

1 2 3 4 5 6   1 2 3 4 5 6

H1 H2 H3 H4 H5 H6   T1 T2 T3 T4 T5 T6

12 possible results; $6 \times 2 = 12$ branches

4. 15 types

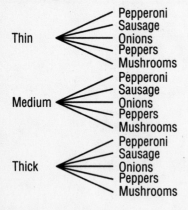

Thin — Pepperoni, Sausage, Onions, Peppers, Mushrooms

Medium — Pepperoni, Sausage, Onions, Peppers, Mushrooms

Thick — Pepperoni, Sausage, Onions, Peppers, Mushrooms

5. 12 choices

6. Answers vary.

7. $\frac{1}{260}$; $26 \times 10 = 260$ ways, 1 correct way

8. 228

9. 288

10. C

11. B

12. 27

Answers for Spiral Review, page 540

1. The vertical scale is broken.

2.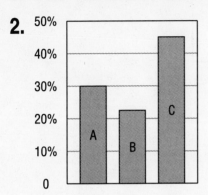

3. Answers vary.

4. $a = 52°, m = 18°, n = 110°,$
$p = 5, k = 6$

5. $a = 145°, b = 65°, c = 55°,$
$d = 95°, s = 12, t = 24, u = 6$

6. $139,728,000 \text{ mi}^2$

7. -7

Answers for Lesson 11.3, pages 545–547

Ongoing Assessment
1. 120　　　　　　　　　**2.** 5040

Practice and Problem Solving

1. Yes

2. No

3. A

4. C

5. B

6. Answers vary.

7. 120

8. 5040

9. 40,320

10. 3,628,800

11. 2; NO, ON; 2

12. 6; <u>CAT</u>, CTA, <u>ACT</u>, ATC, TCA, TAC; 2

13. 24; <u>ARTS</u>, ARST, ASTR, ASRT, ATSR, ATRS, <u>RATS</u>, RAST, RSAT, RSTA, RTAS, RTSA, SART, SATR, SRAT, SRTA, <u>STAR</u>, STRA, <u>TARS</u>, TASR, TRAS, TRSA, <u>TSAR</u>, TSRA; 5 words

14. 24

15. 120

16. 362,880

17. 24

18. $\frac{1}{5040}$

19. $\frac{1}{120}$

20. $\frac{1}{720}$

21. $\frac{1}{5040}$

22. $\frac{1}{40,320}$

23. D

24. B

25. $\frac{1}{6}$

Copyright © McDougal Littell Inc. All rights reserved.

Answers for Lesson 11.4, pages 549–552

Ongoing Assessment

1. $\frac{3}{10}$; You and one of your cousins (but not both) are in 6 of the groups. The probability is $\frac{6}{20}$ or $\frac{3}{10}$.

2. $\frac{9}{20}$; Exactly 2 of the 3 of you are in 9 of these groups. The probability is $\frac{9}{20}$.

Practice and Problem Solving

1.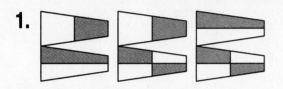

2. Combination

3. Permutation

4.

	A	B	C	D	Combination
1	X	X			{A, B}
2	X		X		{A, C}
3	X			X	{A, D}
4		X	X		{B, C}
5		X		X	{B, D}
6			X	X	{C, D}

6 combinations

(continued)

5.

	A	B	C	D	E	Combination
1	X	X	X			{A, B, C}
2	X	X		X		{A, B, D}
3	X	X			X	{A, B, E}
4	X		X	X		{A, C, D}
5	X		X		X	{A, C, E}
6	X			X	X	{A, D, E}
7		X	X	X		{B, C, D}
8		X	X		X	{B, C, E}
9		X		X	X	{B, D, E}
10			X	X	X	{C, D, E}

10 combinations

6.

	A	B	C	D	E	F	Combination
1	X	X	X	X			{A, B, C, D}
2	X	X	X		X		{A, B, C, E}
3	X	X	X			X	{A, B, C, F}
4	X	X		X	X		{A, B, D, E}
5	X	X		X		X	{A, B, D, F}
6	X	X			X	X	{A, B, E, F}
7	X		X	X	X		{A, C, D, E}
8	X		X	X		X	{A, C, D, F}
9	X		X		X	X	{A, C, E, F}
10	X			X	X	X	{A, D, E, F}
11		X	X	X	X		{B, C, D, E}
12		X	X	X		X	{B, C, D, F}
13		X	X		X	X	{B, C, E, F}
14		X		X	X	X	{B, D, E, F}
15			X	X	X	X	{C, D, E, F}

15 combinations

(continued)

Answers for Lesson 11.4, pages 549–552 (cont.)

7. {1, 3}, {1, 5}, {3, 5}

8. {2, 4, 6}

9. {1, 2, 3, 4}, {1, 2, 3, 5}, {1, 2, 3, 6}, {1, 2, 4, 5}, {1, 2, 4, 6}, {1, 2, 5, 6}, {1, 3, 4, 5}, {1, 3, 4, 6}, {1, 3, 5, 6}, {1, 4, 5, 6}, {2, 3, 4, 5}, {2, 3, 4, 6}, {2, 3, 5, 6}, {2, 4, 5, 6}, {3, 4, 5, 6}

10. {1, 2, 3, 4, 5}, {1, 2, 3, 4, 6}, {1, 2, 3, 5, 6}, {1, 2, 4, 5, 6}, {1, 3, 4, 5, 6}, {2, 3, 4, 5, 6}

11. 3

12. 10

13. 6

14. {R, S}, {R, T}, {S, T}

15. {J, K}, {J, L}, {J, M}, {K, L}, {K, M}, {L, M}

16. {J, K, L}, {J, K, M}, {J, L, M}, {K, L, M}

17. 5

18. Combinations, 35

19. Permutations; 120

20. Permutations; 6

21. Combinations; 15

22. A

23. C

24. a. $\frac{1}{6}$
   b. $\frac{1}{2}$

Answers for Spiral Review, page 552

1. 6 fringetail and 2 popeye

2. $2 \cdot 2 \cdot 3 \cdot 7$

3. $3 \cdot 5 \cdot 13$

4. $2 \cdot 2 \cdot 2 \cdot 17$

5. $2 \cdot 11 \cdot 11$

6. 390 cm^2

7. 384 in.^{2}

8. 251.2 m^2

9. 12

10. −9

11. 9

12. −45

13. 4

14. 9

Answers for Mid-Chapter Assessment, page 553

1. 24

2. 720

3. 5040

4. 362,880

5. Answers vary but could include ways to arrange 10 items or line up 10 people.

6. *Sample answer:* A permutation is an arrangement or listing of objects in which order is important.

7. $\frac{1}{16}$

8. $\frac{1}{16}$

9. $\frac{1}{4}$

10. 16

11. 64,000

12.

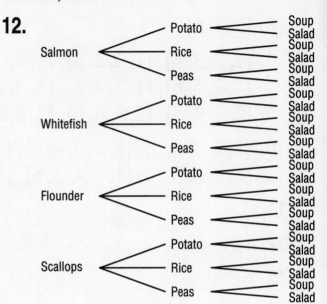

13. 24

14. $\frac{1}{24}$

15. 18

16. 8

Ongoing Assessment

1.–2. Answers vary.

Practice and Problem Solving

1. Fair. The expected value for both players is 1.

2. Not fair. The expected value for Player A is 2.
   The expected value for Player B is 3.

3. Fair. The expected value for both players is $\frac{20}{9}$.

4.

	1	2	3	4	5	6
1	1	2	3	4	5	6
2	2	4	6	8	10	12
3	3	6	9	12	15	18
4	4	8	12	16	20	24
5	5	10	15	20	25	30
6	6	12	18	24	30	36

Expected value for Player A $= \frac{1}{2}$; expected value for Player B $= \frac{3}{4}$; the game is not fair; it could be made fair if Player A gets 3 points for odds and Player B gets 1 point for evens.

5. Yes

6. No; red

7. No; orange

8. 15; $50 \cdot \frac{1}{10} + 25 \cdot \frac{1}{5} + 10 \cdot \frac{3}{10} + 5 \cdot \frac{2}{5} = 15$

9. Answers vary.

10. Player A: 1 point
    Player B: 3 points

11. 1.114; 0.484

12. $.17

13. B

14. Not fair; Player A: $\frac{8}{9}$, Player B: $\frac{10}{9}$; it is fair if
    Player A gets 5 points and Player B gets 4 points.

Answers for Lesson 11.6, pages 561–564

Ongoing Assessment

1. In 1993, the probability that a vehicle had an accident was $\frac{32.8}{203.6} \approx 0.161$. You can predict that your company would cover $0.161 \cdot 150{,}000$ or $24{,}150$ accidents. In 1993, the cost per accident was $\frac{104{,}100}{32.8} \approx \$3174$. You can predict the cost to be $3174 \cdot 24{,}150$ or about $\$76.7$ million.

Practice and Problem Solving

1. $\frac{2}{5}$

2. 60

3. **a.** 28.12 pounds per person
   b. About 878 million

4. About 50

5. About 5 or 6

6. About 75

7. $\frac{2}{5}$ or 0.4

8. About 40,000

9. About 81,000

10. Answers vary.

11. Answers vary.

12. 20

13. Answers vary.

14. Answers vary.

15. 0.37

16. 55.5 million tons

17. C

18. B

19. **a.** 0.215
    b. Answers vary.

Answers for Spiral Review, page 564

1. 83

2. 38

3. 98.8°, 99°, 88°

4. $17.50

5. 21.98 ft, 38.5 ft^2

6. 7.85 m, 4.9 m^2

Answers for Communicating About Mathematics, page 565

1. $\frac{1}{2}$; Alex picks 1 out of 2 keys.

2. $\frac{1}{5}$; Alex picks 1 out of 5 items.

3. 10

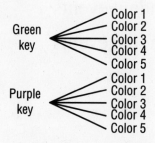

4. $\frac{1}{10}$; there are 10 possible outcomes. Only 1 outcome is correct.

5. Alex would be right 16 out of the 20 times.

6. Answers vary.

Answers for Lesson 11.7, pages 567–569

Ongoing Assessment

1. $\frac{1}{2}$ **2.** $\frac{3}{16}$

Practice and Problem Solving

1. D

2. A

3. C

4. B

5.

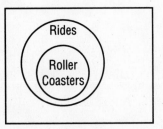

6.

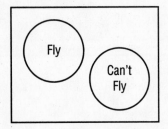

7.

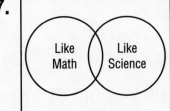

8. Independent events

9. Independent events

10. $\frac{1}{10}$

11. $\frac{1}{10}$

12. $\frac{1}{6}$

13. $\frac{1}{5}$

14.–17. *Sample answers*

14. Some people like only dogs, some people like only cats, and some people like both dogs and cats.

15. People who play musical instruments like music.

16. Either you can swim or you can't swim.

17. Some bowlers own their ball, some bowlers own their shoes, some bowlers own both their ball and their shoes, and some bowlers own neither their ball nor their shoes.

(continued)

18. $\frac{1}{10}$; $\frac{3}{13}$

19. C

20. C

21. a.

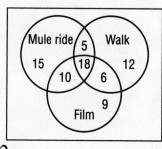

b. $\frac{2}{25}$

Answers for Chapter Review, pages 571–573

1. $\frac{5}{36}$ 　 1, 1 　 1, 2 　 1, 3 　 1, 4 　 1, 5 　 1, 6
　　　　　 2, 1 　 2, 2 　 2, 3 　 2, 4 　 2, 5 　 2, 6
　　　　　 3, 1 　 3, 2 　 3, 3 　 3, 4 　 3, 5 　 3, 6
　　　　　 4, 1 　 4, 2 　 4, 3 　 4, 4 　 4, 5 　 4, 6
　　　　　 5, 1 　 5, 2 　 5, 3 　 5, 4 　 5, 5 　 5, 6
　　　　　 6, 1 　 6, 2 　 6, 3 　 6, 4 　 6, 5 　 6, 6

2. *Sample answer:*

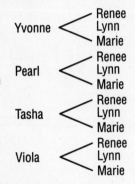

3. 12

Yvonne — Renee, Lynn, Marie

Pearl — Renee, Lynn, Marie

Tasha — Renee, Lynn, Marie

Viola — Renee, Lynn, Marie

4. 60

5. 720

6. $\frac{1}{720}$

7. 10

8. 15

9. Player A: $1\frac{2}{3}$ points

Player B: $2\frac{1}{2}$ points;

not fair because expected values are not equal.

10. Player C: 1 point

Player D: 2 points

11. 75

12. About 8 or 9

13. $\frac{2}{15}$

14. $\frac{7}{30}$

Answers for Chapter Assessment, page 574

1. $\frac{2}{15}$

2. $\frac{13}{15}$

3. 8

4. 8

5. 24

6. Yellow: $\frac{1}{8}$, Blue: $\frac{1}{4}$, Red: $\frac{5}{8}$; Yellow: 5, Red: 5, Blue: 5; fair

7. 736 students

8. Answers vary.

9. $\frac{1}{5}$

10. $\frac{1}{10}$

Answers for Standardized Test Practice, page 575

1. B

2. C

3. C

4. B

5. D

6. B

7. A

8. C

Answers for 12.1, pages 581–583

Ongoing Assessment

1.   3 is added to m to get 2.
Subtract 3 from 2 to get -1.
The solution is $m = -1$.

2.   5 is subtracted from x to get -2.
Add 5 to -2 to get 3.
The solution is $x = 3$.

3. n is multiplied by 5 to get 30.
Divide 30 by 5 to get 6.
The solution is $n = 6$.

4. p is divided by 8 to get 4.
Multiply 4 by 8 to get 32.
The solution is $p = 32$.

Practice and Problem Solving

1. $-3 + 9 = 6, 6 - 9 = -3$

2. $-3 - 4 = -7, -7 + 4 = -3$

3. $-8 \div (-4) = 2, 2 \times (-4) = -8$

4. $3 \times (-2) = -6, -6 \div (-2) = 3$

5. $y + 7 = 0$

6. $m \div (-5) = 2$

7. $n - 8 = -10$

(continued)

222 *Answer Masters*   *Passport to Mathematics Book 2*

8. $t \cdot (-2) = -2$

9.

$-18 \div (-3) = 6, 6$

10. Press the stop button. Take the CD out of the CD player, then place it into the jewel case. Turn off the stereo system, then put the CD away.

11. Subtract 12.

12. Add -3.

13. Divide by -4.

14. Multiply by 10.

15. $, s = 9$

16. $, x = 10$

17. $, y = -11$

18. $, b = 8.3$

19. $, t = \frac{2}{3}$

20. $, m = -1$

21. $, n = -9$

22. $, z = -24$

23.

-6

24.

-7

25.

4.5

(continued)

Answers for 12.1, pages 581–583 (cont.)

26. $\boxed{k} \xrightarrow{-(-4)} \boxed{10}$

6

27. $\boxed{b} \xrightarrow{\times(-3)} \boxed{42}$

-14

28. $\boxed{t} \xrightarrow{\div 3} \boxed{2}$

6

29. $\boxed{n} \xrightarrow{\div 1} \boxed{-2}$

-2

30. $\boxed{a} \xrightarrow{\times(-6)} \boxed{-54}$

9

31. Yes

32. Yes

33. No

34. Yes

35. No

36. No

37. $b + 10 = 6, -4$

38. $m - (-2) = -12, -14$

39. $y \div \frac{1}{4} = 8, 2$

40. $c \times (-3) = -24, 8$

41. To get to the cheese: The mouse should enter the maze, then go left at the first intersection, then continue past a second intersection to the cheese. To get back out of the maze: The mouse should go past the first intersection, then turn right at the second intersection, then continue out of the maze.

42. Ride east $\frac{1}{2}$ block, turn north and ride 1 block, turn east and ride 3 blocks, then turn south and ride 5 blocks.

43. C

44. B

45. 50 marks

Answers for 12.2, pages 585–587

Ongoing Assessment

1. As a subtraction equation

$$m + (-6) = -10 \qquad \text{Write original equation.}$$
$$m - 6 = -10 \qquad \text{Write as subtraction equation.}$$
$$m - 6 + 6 = -10 + 6 \qquad \text{Add 6 to each side.}$$
$$m = -4 \qquad \text{Solution: } m \text{ is by itself.}$$

2. As an addition equation

$$-4 = c - (-5) \qquad \text{Write original equation.}$$
$$-4 = c + 5 \qquad \text{Write as addition equation.}$$
$$-4 - 5 = c + 5 - 5 \qquad \text{Subtract 5 from each side.}$$
$$-9 = c \qquad \text{Solution: } c \text{ is by itself.}$$

Practice and Problem Solving

1. $(-8), (-8); -1$

2. $(-7), (-7); 17$

3. C

4. A

5. D

6. B

7.
$$x - (-21) = 17$$
$$x - (-21) + (-21) = 17 + (-21)$$
$$x = -4$$

8.
$$t - 6 = 13$$
$$t - 6 + 6 = 13 + 6$$
$$t = 19$$

9. Yes

10. No

(continued)

Passport to Mathematics Book 2

Answer Masters **225**

Answers for 12.2, pages 585–587 (cont.)

11. Yes

12. $d + 3 = 10; 7$

13. $-18 = x - 5; -13$

14. $-13 = b + 1; -14$

15. $x + 3 + 5 = 14; 6$ units

16. $x + 90 + 135 = 360; 135°$

17. $x + 35 + 90 = 180; 55°$

18. 10

19. 2.4

20. 14

21. 8

22. 3

23. $\frac{3}{2}$ or $1\frac{1}{2}$

24. Answers vary.

25. C; 123 electoral votes

26. A; 168 electoral votes

27. B; 291 electoral votes

28. D

29. A

30. 19,340 ft

Answers for Spiral Review, page 588

1. a. 56.25, 58.5, 39

b.

```
       ┌─────────┬──────────────┐
  •────┤         │              ├────•
       └─────────┴──────────────┘
  32   39        58.5     71.5   80
```

2. 42

3. 2.5%

4. $A(1, 2), B(2, 5), C(4, 3), D(6, 4), E(3, 1)$

5. $A(-2, 4), B(-1, 7), C(1, 5) D(3, 6), E(0, 3)$

6. 1 unit to the left and 2 units down

Answers for 12.3, pages 593–596

Ongoing Assessment

1. $23.40 gain

2. Answers vary.

Practice and Problem Solving

1. $-4m = 32, -8$

2. $6b = -78, -13$

3. $-3y = -51, 17$

4. C

5. B

6. D

7. A

8. -12; the mean represents the average number of points missed per test.

9. Negative, $-3\frac{2}{5}$

10. Positive, 13

11. Positive, 18

12. Negative, $-9\frac{1}{2}$

13. About 50; $50\frac{1}{2}$

14. About -3; $-3\frac{1}{10}$

15. About -8; $-7\frac{1}{2}$

16. About $\frac{1}{2}$; $\frac{3}{5}$

17. About -10; -10.5

18. About -16; -15

19. About 7; 7.24

20. About 0.1; 0.125

21. About $\frac{1}{6}$; $\frac{1}{6}$

22. 4

23. -12

24. 2

25. 0

26. -2.4

27. $\frac{3}{4}$

28. $1\frac{1}{5}$

29. -3

30. -12

31. $7\frac{1}{2}$ yards

32. $-5n = 0; 0$

33. $-22 = -8n; 2\frac{3}{4}$

34. $-6 = 9n; -\frac{2}{3}$

35. $-7n = 91; -13$

36. $9x = 135; 15$ in.

(continued)

37. $16x = 120$; 7.5 ft

38. $3x = 2.4$; 0.8 m

39. $-8, -4, -2$; as the coefficient of d doubles, d is halved.

40. $-2, -4, -6$; as the product decreases by 3, x decreases by 2.

41. $A = 20N$; $1000 = 20 \times 50$

42. $T = 2S$; $500 = 2 \times 250$

43. $S = 2.5C$; $250 = 2.5 \times 100$

44. $F = \frac{1}{5}A$; $200 = \frac{1}{5} \times 1000$

45. C

46. A

47. 192 chicken wings

6 cups soy sauce

6 cups water

24 tablespoons honey

48 cloves garlic, crushed

24 tablespoons chopped fresh ginger

Each ingredient must be multiplied by 24.

Answers for Spiral Review, page 596

1. 108 cm, 544 cm^2

2. 30 ft, 30 ft^2

3. 69.08 cm, 379.94 m^2

4. 410 in.^{2}

5. $10{,}110.8$ mm^2

6. 246 ft^2

7. A loss of \$430

Answers for Mid-Chapter Assessment, page 597

1. c; -9

2. a; -4

3. d; -36

4. b; -15

5. For steam to turn into water, the temperature must fall below $212°$F. For water to turn into ice, the temperature must fall to at least $32°$F.

6. $n - 3 = -4;\ -1$

7. $-5 = x + 4;\ -9$

8. $m + 3 = 6;\ 3$

9. You should add -4, not 4.
$$n - (-4) = -6$$
$$n - (-4) + (-4) = -6 + (-4)$$
$$n = -10$$

10. You should add 5 instead of subtracting 6.
$$6 = x - 5$$
$$6 + 5 = x - 5 + 5$$
$$11 = x$$

11. $-\frac{15}{3} = -5$, not 5.
$$3p = -15$$
$$\frac{3p}{3} = -\frac{15}{3}$$
$$p = -5$$

12. -3

13. -3

14. $4\frac{1}{2}$

15. $6x = -4780$; loss is \$796.67.

Ongoing Assessment

1. About 309 pounds

2. Set up a proportion similar to Example 2.

Practice and Problem Solving

1. multiply

2. divide

3. C; 18.89 gallons

4. A; $6300

5. B; 0.24 gallons

6. You cannot divide by 0.

7. 120

8. 252

9. $\frac{5}{4}$ or $1\frac{1}{4}$

10. 120

11. 11.27

12. $\frac{15}{32}$

13. -56.1

14. -30.34

15. 263.25 calories

16. $\frac{x}{12} = 15$ or $\frac{x}{12} = 15$; 180 cm^2

17. $\frac{x}{13} = 13$; 169 yd^2

18. $\frac{x}{2.5} = 2.6$ or $\frac{x}{2.6} = 2.5$; 6.5 m^2

19. Negative

20. Positive

21. Negative

22. Positive

23. 10, 20, 30; as the quotient increases by 2, s increases by 10.

24. $-4, -5, -6$; as the quotient decreases by 5, r decreases by 1.

25. 5.94, 6.93, 7.92; as the quotient increases by 0.09, p increases by 0.99.

(continued)

26. B

27. A

28. Height of souvenir flag: 4 in.
Width of yellow stripe: 2 in.
Width of red and blue stripes: 1 in.

Answers for 12.5, pages 605–608

Ongoing Assessment

1.

$$2x + 5 = 13$$ Write original equation.

$$2x + 5 - 5 = 13 - 5$$ Subtract 5 from each side.

$$2x = 8$$ Simplify.

$$\frac{2x}{2} = \frac{8}{2}$$ Divide each side by 2.

$$x = 4$$ Solution: x is by itself.

2.

$$-4 + \frac{n}{4} = 4$$ Write original equation.

$$-4 + \frac{n}{4} + 4 = 4 + 4$$ Add 4 to each side.

$$\frac{n}{4} = 8$$ Simplify.

$$4 \cdot \frac{n}{4} = 4 \cdot 8$$ Multiply each side by 4.

$$n = 32$$ Solution: n is by itself.

3.

$$-2y - 6 = 0$$ Write original equation.

$$-2y - 6 + 6 = 0 + 6$$ Add 6 to each side.

$$-2y = 6$$ Simplify.

$$\frac{-2y}{-2} = \frac{6}{-2}$$ Divide each side by –2.

$$y = -3$$ Solution: y is by itself.

Practice and Problem Solving

1. Write original equation; subtract 6 from each side; simplify; multiply each side by 2; solution: n is by itself.

2. Write original equation; add 13 to each side; simplify; divide each side by 9; solution: y is by itself.

(continued)

3. $-2n + 5 = -7; 6$

4. 5

5. Subtract 10 from each side.

6. Add 16 to each side.

7. Add 5 to each side.

8. $2t - 8 = 14; 11; 22$

9. $-5b + 3 = 28; -5; 25$

10. $\frac{m}{3} + 1 = -9; -30; -10$

11. -4

12. -3

13. 21

14. 0

15. -10

16. 2

17. -1

18. -5

19. $5\frac{1}{2}$

20. $57°, 33°$

21. $56°, 95°, 29°$

22. $130°, 50°$

23. $a = -6, b = 2,$
$c = -1, d = 8$

24. Swimmers Club; because you would get more time for your money.

25. Verbal Model:

Amount per subscriber		Number of subscribers		Amount from company		Amount of goal
	$\cdot$		$+$		$=$	

Labels: Amount per subscriber $= \$1.25$
Number of subscribers $= n$
Amount from company $= \$175$
Amount of goal $= \$600$

Equation: $1.25n + 175 = 600$
$n = 340$

26. D

27. B

(continued)

Answers for 12.5, pages 605–608 (cont.)

28. a. Women's: $10\frac{5}{12}$ in.

Men's: $10\frac{137}{150}$ in.

b. Women's shoe size: 1, length of foot: $8\frac{5}{12}$ in.

Men's shoe size: $5\frac{1}{2}$, length of foot: 10.08 in.

c. $\frac{1}{3}\left(\boxed{\begin{matrix}\text{Shoe}\\\text{size}\end{matrix}} - \boxed{\begin{matrix}\text{Smallest}\\\text{shoe size}\end{matrix}} \right) = \boxed{\begin{matrix}\text{Foot}\\\text{length}\end{matrix}} - \boxed{\begin{matrix}\text{Smallest}\\\text{foot length}\end{matrix}}$

Answers for Spiral Review, page 608

1. $9 + x = -14$; -23

2. $-7 = x - 15$; 8

3. $42 = 3\frac{1}{2} \cdot x$; 12

4. $-24 = \frac{1}{4}x$; -96

5.

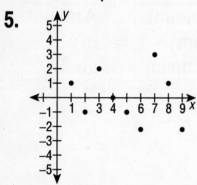

Quadrants I and IV

6. 3

7. 48

8. 5

9. 14

10. 70

Answers for Communicating About Mathematics, page 609

1. Determine the number of years between 4 and 11 and multiply by $\frac{1}{3}''$.

   $x = \frac{1}{3}(11 - 4)$

   $x = \frac{1}{3}(7)$

   $x = \frac{7}{3}$ or $2\frac{1}{3}$ in.

2. Infants': $\frac{1}{3}(\text{shoe size} - 0) = \text{foot length} - 3\frac{11}{12}$

   Girls': $\frac{1}{3}(\text{shoe size} - 7) = \text{foot length} - 5\frac{11}{12}$

   Boys': $\frac{1}{3}(\text{shoe size} - 8) = \text{foot length} - 6\frac{5}{12}$

   Men's: $\frac{1}{3}(\text{shoe size} - 5\frac{1}{2}) = \text{foot length} - 10.08$

   Women's: $\frac{1}{3}(\text{shoe size} - 1) = \text{foot length} - 8\frac{5}{12}$

3. Each equation uses $\frac{1}{3}$, shoe size, and foot length. Each equation has a different minimum shoe size and a different minimum foot length.

4. No

5. Bob Lanier's feet are 15.58 in. long.

Answers for 12.6, pages 613–615

Ongoing Assessment

1.

Blocks in top row, input x	1	2	3	4	5
Perimeter, output P	12	14	16	18	20

Each time x increases by 1, the perimeter increases by 2.

2. $P = 10 + 2x$.

Practice and Problem Solving

1. B

2. C

3. A

4. 63, 71, 79, 87, 95, 103

5. $C = 75 + 4x$

Input, x	1	2	3	4	5	6
Output, C	79	83	87	91	95	99

6. Maple Creek; the cost is $4 less for 6 people.

7.

x	1	2	3	4	5	6
y	7	12	17	22	27	32

8.

x	1	2	3	4	5	6
y	$2\frac{1}{3}$	$2\frac{2}{3}$	3	$3\frac{1}{3}$	$3\frac{2}{3}$	4

9.

x	1	2	3	4	5	6
y	-4	-7	-10	-13	-16	-19

10.

x	1	2	3	4	5	6
y	0	5	10	15	20	25

(continued)

11.

x	1	2	3	4	5	6
y	$1\frac{1}{6}$	$1\frac{1}{3}$	$1\frac{1}{2}$	$1\frac{2}{3}$	$1\frac{5}{6}$	2

12.

x	1	2	3	4	5	6
y	-2.5	0	2.5	5	7.5	10

13. $y = 16 - 2x$ **15.** $y = 3x + 5$

14. $y = 3x + 3$ **16.** $y = 2x + 0.5$

17.

t (seconds)	1	2	3	4
d (miles)	0.2	0.4	0.6	0.8

18. $d = 0.2t$

19.

x (units)	1	2	3	4	5
V (cubic units)	3	6	9	12	15

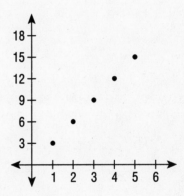

20.

x (units)	1	2	3	4	5
V (cubic units)	0.79	1.57	2.36	3.14	3.93

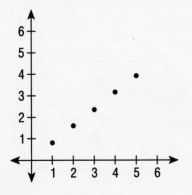

(continued)

21. C;

x (in.)	1	2	3	4	5
S (in.^{2})	6	24	54	96	150

22. C

23. B

24. 4.5 mm, 2.9 mm

Answers for Chapter Review, pages 617 and 618

1. 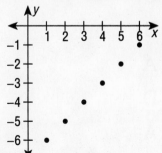 , −8

2. , 5

3. , 14

4. , 26

5. 8

6. −3

7. −7

8. −29

9. −5

10. 15.5

11. 50

12. 7.5

13. 62.5

14. 4

15. −45.5

16. 21

17. 12

18. $\frac{4}{5}$

19. 3

20. 30

21.

x	0	1	2	3	4	5	6
y	−3	−1	1	3	5	7	9

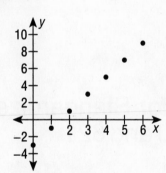

22. When $x = 1$, $y = -6$. For every 1 unit increase in x, there is a 1 unit increase in y. $y = x - 7$

Answers for Chapter Assessment, page 619

1. $b - 4 = -3,$

$b = 1$

2. $\dfrac{b}{6} = 6.5,$

$b = 39$

3. $-2b - 7 = 13,$

$b = -10$

4. -25

5. -5

6. -30

7. Positive; the equation is of the form (Negative)(Positive) = Negative.

8. Negative; the equation is of the form (Negative)(Negative) = Positive.

9. Negative; the equation is of the form $\dfrac{\text{Negative}}{\text{Positive}} = $ Negative.

10. $9x = 117$; 13 cm

11. $5 + 8 + x = 22$; 9 ft

12. $y = 16 - 3x$

13. $y = 4x + 1$

14. 10

15. 30%

Answers for Standardized Test Practice, pages 620 and 621

1. C	8. C	15. D
2. A	9. B	16. D
3. D	10. B	17. C
4. D	11. C	18. B
5. B	12. D	19. A
6. D	13. B	20. D
7. C	14. B	21. A

240 *Answer Masters*

Passport to Mathematics Book 2

Answers for Cumulative Review, page 622

1. 156.25
2. 105
3. 40
4. 84
5. 34.48 m
6. 9.90 in.
7. 13 mm
8. 288 in.^{3}; 312 in.^{2}
9. 321.9 ft^3; 285.7 ft^2
10. 2210.6 cm^3; 954.6 cm^2
11. −13
12. −3
13. 5
14. −28
15. 9
16. 14
17. −32
18. 21
19. $\frac{1}{2}$
20. $\frac{2}{9}$

21. $\frac{5}{6}$
22. 1
23. −3
24. +6
25. ÷12
26. ×8
27. +(−7)
28. −(−10)
29. ×(−2)
30. ÷(−9)
31. $\times \left(-\frac{4}{3} \right)$
32. −2
33. −19
34. 6
35. 9
36. −15
37. −64
38. 14
39. 7
40. −4